# SpringerBriefs in Electrical and Computer Engineering

More information about this series at http://www.springer.com/series/10059

Saleh Faruque

# Radio Frequency Multiple Access Techniques Made Easy

 Springer

Saleh Faruque
Department of Electrical Engineering
University of North Dakota
Grand Forks, ND, USA

ISSN 2191-8112       ISSN 2191-8120   (electronic)
SpringerBriefs in Electrical and Computer Engineering
ISBN 978-3-319-91649-1       ISBN 978-3-319-91651-4   (eBook)
https://doi.org/10.1007/978-3-319-91651-4

Library of Congress Control Number: 2018948162

Printed on acid-free paper

This Springer imprint is published by the registered company Springer Nature Switzerland AG
The registered company address is: Gewerbestrasse 11, 6330 Cham, Switzerland

# Preface

In telecommunications, multiple access techniques enable many users to share the same spectrum in the frequency domain, time domain, code domain, or phase domain. It begins with a frequency band, allocated by the Federal Communications Commission (FCC). FCC provides licenses to operate wireless communication systems over given bands of frequencies. These bands of frequencies are further divided into smaller bands (channels) and reused to provide services to other users. This is governed by the International Telecommunication Union (ITU). ITU generates standards such as FDMA, TDMA, CDMA, OFDMA, etc. for wireless communications.

As the size and speed of digital data networks continue to expand, bandwidth efficiency becomes increasingly important. This is especially true for broadband communications, where the choice of modulation schemes is important, keeping in mind the available bandwidth resources allocated by FCC. With these constraints in mind, this book will present a comprehensive overview of multiple access techniques used in the cellular industry. Numerous illustrations will be used to bring students up-to-date in key concepts, underlying principles, and practical applications of multiple access techniques so that they can readily put them into work in the industry.

The list of topics under consideration is presented below:

- Introduction
- Mode of Operation: FDD, TDD
- FDMA Technique
- TDMA Technique
- CDMA Technique
- OFDMA Technique
- Conclusions

This text has been primarily designed for electrical engineering students in the area of telecommunications. However, engineers and designers working in the area of wireless communications would also find this text useful. It is assumed that the reader is familiar with the general theory of telecommunications.

In closing, I would like to say a few words about how this book was conceived. It came out of my long industrial and academic career. During my teaching tenure at the University of North Dakota (UND), I developed a number of graduate-level elective courses in the area of telecommunications, which combine theory and practice. This book is a collection of my courseware, research activities, and hands-on experience in wireless communications.

I am grateful to the UND and the School for the Blind, North Dakota, for affording me this opportunity. This book would never have seen the light of day had UND and the State of North Dakota not provided me with the technology to do so. My heartfelt salute goes out to the dedicated developers of these technologies, who have enabled me and others visually impaired to work comfortably.

I would like to thank my beloved wife, Yasmin, an English literature buff and a writer herself, for being by my side throughout the writing of this book and for patiently proofreading it and to my darling son, Shams, an electrical engineer himself, for providing technical support in formulation, simulation, and experimentation when I needed it. For this, he deserves my heartfelt thanks.

In spite of all this support, there may still be some errors in this book. I hope that my readers forgive me for them. I shall be amply rewarded if they still find this book useful.

Grand Forks, ND, USA                                          Saleh Faruque

# Contents

# Contents

# Chapter 1
# Introduction

**Abstract** Multiple access techniques enable many users to share the same spectrum in the frequency domain, time domain, code domain, or phase domain. It begins with a frequency band, allocated by FCC (Federal Communication Commission). FCC provides licenses to operate wireless communication systems over given bands of frequencies. These bands of frequencies are finite and have to be further divided into smaller bands (Channels) and reused to provide services to other users. This is governed by the International Telecommunication Union (ITU). ITU generates standards such as FDMA, TDMA, CDMA, OFDMA, etc. for wireless communications. These standards along with numerous illustrations are presented to bring the students up-to-date in key concepts, underlying principles, and practical applications of multiple access techniques so that they can readily put them into work in the cellular industry.

## 1.1 Introduction to Multiple Access Techniques

Multiple access technique is well known in cellular communications [1–7]. It enables many users to share the same spectrum in the frequency domain, time domain, code domain, or phase domain. It begins with a frequency band, allocated by FCC (Federal Communication Commission). FCC provides licenses to operate wireless communication systems over given bands of frequencies. These bands of frequencies are finite and have to be further divided into smaller bands (Channels) and reused to provide services to other users. This is governed by the International Telecommunication Union (ITU). ITU generates standards such as FDMA, TDMA, CDMA, OFDMA, etc. for wireless communications. Figure 1.1 illustrates the basic concept of various multiple access techniques currently in use.

*In FDMA* (frequency division multiple access) [1–3], a carrier frequency is assigned to a single user. In this scheme, the channel is occupied by a single user for the entire duration of the call. FDMA is the multiple access technique used in AMPS (advanced mobile phone system), also known as the first-generation (1G) cellular communication system.

S. Faruque, *Radio Frequency Multiple Access Techniques Made Easy*,
SpringerBriefs in Electrical and Computer Engineering,
https://doi.org/10.1007/978-3-319-91651-4_1

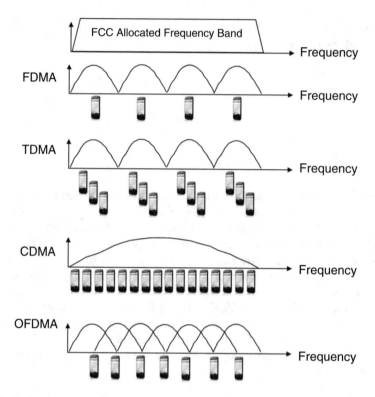

**Fig. 1.1** Basic concept of multiple access techniques

*In TDMA* (time division multiple access) [1–3], a carrier frequency is assigned to multiple users. In this scheme, the channel is used by multiple users, one at a time as depicted in Fig. 1.1 [4–8]. TDMA is the multiple access technique used in North American TDMA and European GSM wireless standards. TDMA is also known as the second-generation (2G) cellular communication system.

*In CDMA* (code division multiple access) [1–3], multiple users have access to the same carrier frequency at the same time. CDMA is also known as spread spectrum technique. This is accomplished by spreading the data by means of orthogonal codes. In CDMA, each user is assigned a unique orthogonal code, while each user uses the same carrier frequency. The use of orthogonal codes enables each user to communicate without interference. CDMA is used in IS95 standard.

*OFDMA* (orthogonal frequency division multiple access) [6, 7] is an extension of FDMA, where each band is placed at the null of the adjacent band, so that adjacent bands are orthogonal to each other and do not interfere. OFDMA is the multiple access technique used in WiMAX (Worldwide Interoperability for Microwave Access) and LTE (Long-Term Evolution) protocol. It may be noted that WiMAX is an IEEE 802.16 standard while LTE is a standard developed by the 3GPP group. Both standards are surprisingly similar and bandwidth efficient. OFDMA is used in 4G cellular standard.

In any multiple access techniques, multiple users have access to the same spectrum, so that the occupied bandwidth does not exceed the FCC allocated channel. Furthermore, as the size and speed of digital data networks continue to expand, bandwidth efficiency becomes increasingly important. This is especially true for broadband communication, where the choice of modulation scheme is important keeping in mind the available bandwidth resources, allocated by FCC. With these constraints in mind, this book will present a comprehensive overview of multiple access techniques used in the cellular industry. The concept of frequency reuse will be studied along with carrier to interference (C/I). The classical cell reuse plan will be studied next with examples of various frequency plans related to OMNI and Sectorization schemes. Numerous illustrations will be used to bring students up to date in key concepts, underlying principles, and practical applications of multiple access techniques so that they can readily put them into work in the industry.

## 1.2 Mode of Operation

### 1.2.1 Background

Cellular communication is a full duplex, or simply a duplex communication system. Duplex is referred to as a two-way communication, where two users can communicate with each other simultaneously. Figure 1.2 illustrates the basic concept of a full duplex cellular communication system, currently in use all over the world.

According to the communication protocol, the cellular base station assigns a carrier frequency to the cell phone. Once the carrier frequency is assigned, the cell phone modulates the carrier frequency by means of voice, data, or video. It then amplifies the modulated carrier frequency and sends it to the antenna for transmission. Upon receiving, the cellular radio at the base station responds to the cell phone by means of a similar communication protocol. The link is maintained in both directions, either in the frequency domain or in the time domain. This is governed by two basic mode of operations listed below:

- Frequency division duplex (FDD)
- Time division duplex (TDD)

A brief description of these communication modes are presented below.

**Fig. 1.2** Illustration of full duplex wireless communication

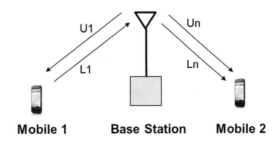

## 1.2.2 Frequency Division Duplex (FDD)

*In FDD*, all the available channels are divided into two bands (lower band and upper band) and grouped as pairs (L1U1, L2U2... LnUn), as depicted in Fig. 1.3. As can be seen, FDD uses two different frequencies, one for the upload and the other for the download, separated by a guard band. As a result, both transmissions can take place at the same time without interference. This scheme is known as FDMA-FDD. A brief description of FDD communication is presented below:

- The base station modulates the carrier frequency (U1) from the upper band and sends the modulated carrier to the mobile. The input modulating signal can be either analog or digital.
- Since the mobile is tuned to the same carrier frequency, it receives the modulated carrier from the base after a propagation delay. It then demodulates the carrier and recovers the information signal.
- In response, the mobile modulates a different carrier frequency (L1) from the lower band and transmits back to the base.
- The base station receives the modulated signal from the mobile, demodulates, and recovers the information.
- The process continues until one of the transmitters terminates the call.

**Fig. 1.3** Frequency division duplex (FDD)

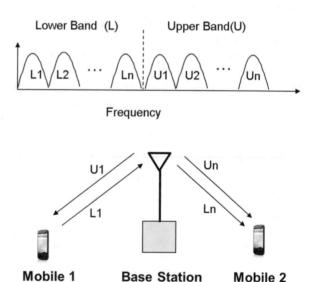

### 1.2.3   Time Division Duplex (TDD)

In TDD, a single frequency is time shared between the uplink and the downlink. The duration of transmission in each direction is generally short, in the order of ms. In this scheme, when the mobile transmits, base station listens, and when the base station transmits, mobile listens. This is accomplished by formatting the data into a "Frame," where the frame is a collection of several time slots. Each time slot is a package of data, representing digitized voice, digitized text, digitized video, and synchronization bits (sync. bits). The sync bits are unique, which is used for frame synchronization. Details will be presented in subsequent *chapters*.

Figure 1.4 illustrates a typical frame and the TDD transmission scheme. According to TDD transmission protocol, both the base station and the mobile use the same carrier frequency.

Figure 1.4 demonstrates a typical TDD scheme.

The transmit/receive mechanism between the base station and mobile are as follows:

- The base station modulates the carrier frequency by means of the digital information bits in frame F(B) and transmits to the mobile.
- Since the mobile is tuned to the same carrier frequency, it receives the frame F (B) after a propagation delay $t_p$.
- Mobile synchronizes the frame using the sync bits and downloads the data.
- After a guard time $t_g$, mobile transmits its own frame F(M) to the base.

**Fig. 1.4**  TDD frame structure. (**a**) Frame. (**b**) Frame transmission scheme

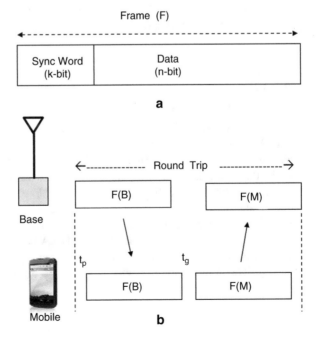

- Base receives the frame F(M) from the mobile after a propagation delay $t_p$ an, maintains sync using the sync bi5ts, and downloads the da5ta.
- A round-trip communication is now complete.
- The communication continues until one terminates the call.

As can be seen, the TDD schemes require a propagation delay and a guard time between transmission and reception. The complete round-trip delay $T_d$ must be sufficient to accommodate the frame, the propagation delay, and the guard time. Therefore, the round-trip delay can be written as

$$T_d = 2(F + t_p + t_g) \tag{1.1}$$

where:

$F$ = frame length
$t_p$ = propagation delay
$t_g$ = guard time

In cellular communications, such as OFDMA and LTE (4G), the traffic in both directions is not balanced. There is scheduling protocol, which can be dynamically controlled to offer high-speed data over the downlink and low-speed data over the uplink. In TDD, this is accomplished by transmitting more time slots over the downlink, thereby supporting more capacity. For these reasons, TDD is used in 4G cellular system as WiMAX and LTE standards. Details will be presented in later chapters.

This introductory chapter will define various multiple access techniques currently in used in the communication industry, identify several challenges, and propose possible solutions. Detailed description, practical realization, and examples will be presented in subsequent chapters.

## 1.3  FDMA Technique

*In FDMA* (frequency division multiple access) (Fig. 1.5), a pair of carrier frequencies are used during a call: one from the lower band and one from the upper band. The lower band frequency is used by the mobile radio, and the upper band frequency is used by the base station radio. In this technique, two channels are occupied during the entire duration of the call. It may be noted that lower frequencies are used in the mobile because propagation decay is logarithmic as a function of frequency. Since the mobile transmit power is much lower than the base station radio, it is necessary to use lower frequencies in the mobile phone. FDMA is the multiple access technique used in AMPS (advanced mobile phone system), also known as the first-generation (1G) cellular communication system.

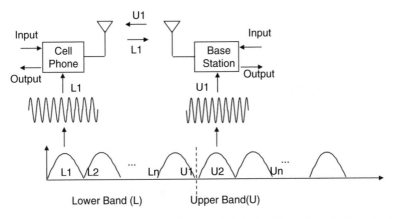

**Fig. 1.5** FDMA technique. A pair of channels are used during a call: one from the lower band and one from the upper band. The lower band frequency is used by the mobile radio, and the upper band frequency is used by the base station radio. Both channels are occupied during the entire duration of the call

## 1.4  TDMA Technique

*TDMA* (time division multiple access) is an extension of FDMA, where each FDMA channel is time shared by multiple users, one at a time. In this technique, a pair of FDMA channel is used during a call: one from the lower band and one from the upper band. The lower band frequency is time shared by several mobiles. The upper band frequency is also time shared synchronously by the base station radio. Both channels are occupied during the entire duration of the call.

The synchronization is achieved by means of a special frame structure, where the frame is a collection of time slots. Each time slot is assigned to a mobile. Figure 1.6 shows the frame structure used in the North American 2G TDMA system. As shown in the figure, there are six time slots in the frame, where two time slots are assigned per user in sequence. This implies that, when one mobile has access to the channel, the other mobiles are idle. Therefore, TDMA synchronization is critical for data recovery and collision avoidance.

TDMA has several advantages over FDMA:

- Increased channel capacity
- Greater immunity to noise and interference
- Secure communication
- More flexibility and control

Moreover, it allows the existing FDMA standard to coexist in the same TDMA platform, sharing the same RF spectrum. The detailed technology, adopted as the North American TDMA standard, will be presented in this book.

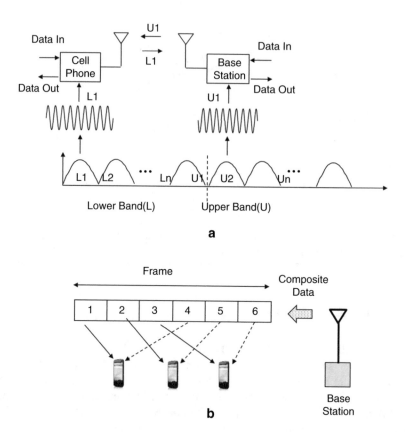

**Fig. 1.6** TDMA technique. It is an extension of FDMA, where each FDMA channel is time shared by several mobiles. Data transmission takes place over a frame

## 1.5  CDMA Technique

*In CDMA* (code division multiple access), multiple users have access to the same carrier frequency at the same time as depicted in Fig. 1.7. Here, a single carrier frequency is used by several users, where each user is assigned a unique orthogonal code.

The use of orthogonal codes enables each user to communicate without interference. CDMA is also known as spread spectrum technique. This is accomplished by multiplying each information bit by means of an n-bit orthogonal code. Multiplication in this process is referred to as exclusive OR (EXOR) operation. For example, when the binary bit 0 is multiplied by a 4 bit orthogonal code 0101, we write

$$0 \, EXOR \, (0101) = 0101 \tag{x1.2}$$

**Fig. 1.7** CDMA technique. A single frequency is used my multiple user

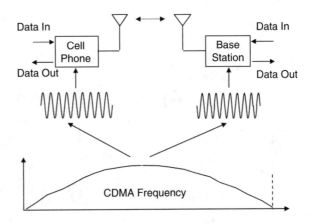

This is the orthogonal code reproduced due to exclusive OR operation. Moreover, the bit rate is also multiplied by a factor of four, thereby spreading the spectrum by a factor of four as well.

Similarly, when the binary bit 1 is multiplied by the same orthogonal code, we obtain

$$1 \text{ EXOR } (0101) = 1010 \tag{x1.3}$$

This is the inverse of the orthogonal code. This code is also known as antipodal code. The spectrum is also spread by a factor of four.

Therefore, when an information bit is multiplied by an n-bit orthogonal code, the spectrum is spread by a factor of n, enabling multiple users to share the same spectrum at the same time. For this reason a large bandwidth is assigned for CDMA radio. Details will be presented in this book as a separate chapter.

Let's examine a CDMA radio based on 4-bit orthogonal code, where the code sequence is given by 0101. This is shown in Figs. 1.8 and 1.9. In Fig. 1.8, the binary bit 0 is multiplied by the 4 bit orthogonal code to reproduce the orthogonal code, which represents the information bit 0. Now, the modulator modulates the carrier frequency by means of the orthogonal code and transmits through the antenna. The receiver intercepts the modulated carrier frequency, demodulates, and recovers the orthogonal code 0101. Since the exclusive OR gate also uses the same orthogonal code, we obtain

$$(0101) \text{ EXOR } (0101) = 0000 \tag{x1.4}$$

This represents the original binary value 0.

Similarly, in Fig. 1.9, the binary bit 1 is multiplied by the same 4 bit orthogonal code to generate the antipodal code 1010, which represents the information bit 1. Now, the modulator modulates the carrier frequency by means of the antipodal code and transmits through the antenna. The receiver intercepts the modulated

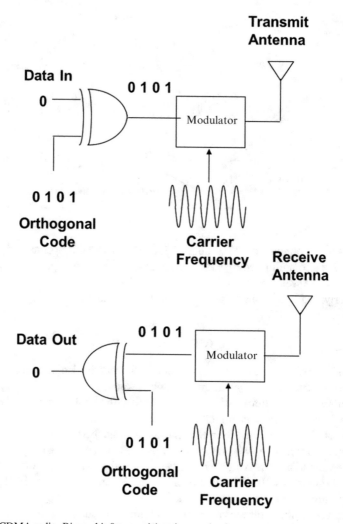

**Fig. 1.8** CDMA radio. Binary bit 0 transmit/receive mechanism

carrier frequency, demodulates, and recovers the antipodal code 1010. Since the exclusive OR gate also uses the same orthogonal code, we obtain

$$(0101) \text{ EXOR } (1010) = 1111 \qquad (1.5)$$

This represents the original binary value 1.

In summary, CDMA is a branch of multiple access technique, where multiple users have access to the same spectrum through orthogonal codes. Orthogonal codes are binary valued and have equal number of 1's and 0's. Therefore, for an n-bit orthogonal code, there are n orthogonal codes. In CDMA, each user is assigned a

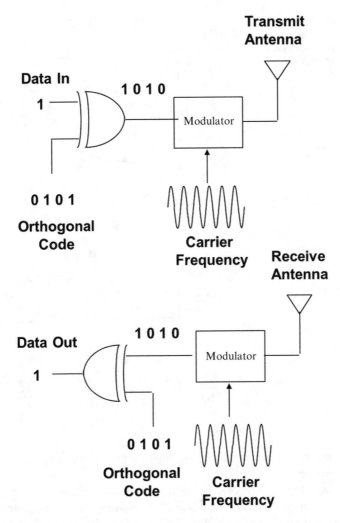

**Fig. 1.9** CDMA radio. Binary bit 1 transmit/receive mechanism

unique orthogonal code. Yet, there is a limitation of using all the codes, which is related to channel capacity. We will address this issue in details in this book.

## 1.6   OFDMA Technique

*OFDMA* (orthogonal frequency division multiple access) is an extension of FDMA, where each channel is placed at the null of the adjacent channel, so that adjacent channels are orthogonal to each other and do not interfere. Figure 1.10

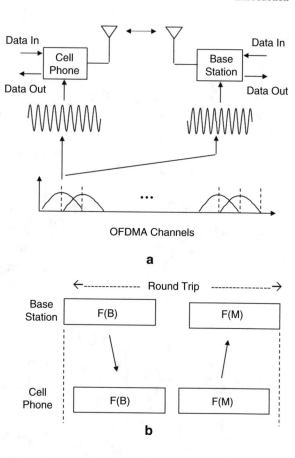

**Fig. 1.10** OFDMA.
(**a**) Each channel is placed at the null of adjacent channels to acquire orthogonal property. (**b**) TDD technique

shows a typical OFDMA technique, operating in the TDD mode, where the same channel is used in both directions.

It may be noted that the orthogonal property is due to power spectral density associated with nonperiodic digital signals. The power spectral density, on the other hand, is derived by Fourier transform. This topic will be further discussed in the OFDMA chapter of this book.

While both FDD and TDD can be used in OFDMA, the TDD technique is more spectrally efficient. TDD can be used to alter the number of frames in each direction. This is a common feature in WiMAX and LTE, where the downlink bit rate is much higher than the uplink. Figure 1.10b shows a typical frame and the TDD transmission scheme. According to TDD transmission protocol, both the base station and the mobile use the same carrier frequency. According to the communication protocol, when the base station transmits, the mobile listens, and when the mobile transmits, the base listens. As a result, there is a round-trip delay which is given by

$$T_d = 2\left(F + t_p + t_g\right) \tag{1.6}$$

where:

$T_d$ = round-trip delay

$F$ = frame length

$t_p$ = propagation delay

$t_g$ = guard time between frames

OFDMA is the multiple access technique used in WiMAX (Worldwide Interoperability for Microwave Access) and LTE (Long-Term Evolution) protocol. WiMAX is an IEEE 802.16 standard, while LTE is a standard developed by the 3GPP group. Both standards are surprisingly similar and bandwidth efficient.

## 1.7    Conclusions

This introductory chapter provides a brief overview of various multiple access techniques currently in use all over the world. The roles of the FCC and the ITU are also discussed. The FCC provides licenses to operate wireless communication systems over given bands of frequencies. These bands of frequencies are finite and have to be further divided into smaller bands (channels) and reused to provide services to other users. This is governed by the International Telecommunication Union (ITU). ITU generates standards such as FDMA, TDMA, CDMA, OFDMA, etc. for wireless communications. These standards are currently in use all over the world. These standards along with numerous illustrations are presented to bring the students up to date in key concepts, underlying principles, and practical applications of multiple access techniques so that they can readily put them into work in the cellular industry. Details are presented in subsequent chapters.

## References

1. IS-54. (1989, December). *Dual-mode mobile station-base station compatibility standard* (EIA/TIA project number 2215).
2. IS-95. (1993, March 15). *Mobile station - Base station compatibility standard for dual mode wide band spread spectrum cellular systems* (TR 45, PN-3115).
3. Mac Donald, V. H. (1979, January). The cellular concept. *The Bell System Technical Journal, 58*(1).
4. FCC, Federal Communication Commission, Washington, DC.
5. ITU, International Telecommunications Union, Paris, France.
6. ITU-R. *Requirements related to technical performance for IMT-advanced radio interface(s)* (Report M.2134), Approved in November 2008.
7. Parkvall, S., & Astely, D. (2009, April). The evolution of LTE toward LTE Advanced. *Journal of Communications, 4*(3), 146–154. https://doi.org/10.4304/jcm.4.3.146-154.
8. Faruque, S., & Semke W. *Phase division multiple access (PDMA) for telecommunications* (Nsf2015 Proposal 1547072).

# Chapter 2
# Simplex, Duplex, FDD, and TDD

**Abstract** In telecommunication, there are various ways to communicate between two points, referred to as mode of operation. This is usually done in three ways: simplex, half-duplex, and full duplex. Simplex is a one-way communication such as broadcasting. Half duplex is a two-way communication but one at a time such as taxi dispatch system. Full duplex is a method of two-way communication, where two users can communicate with each other simultaneously such as land-mobile telephone system. In full duplex communication, the link is maintained in both directions, either in the frequency domain or in the time domain. This is governed by two basic communication schemes: frequency division duplex (FDD) and time division duplex (TDD). A brief description of these communication modes is presented in this chapter.

## 2.1 Introduction

In telecommunication, there are various ways to communicate between two points, referred to as "modus operendi" or simply mode of operation. This is usually done in three ways: simplex, half-duplex, and full duplex. A brief description of these communication modes are presented below [1, 2].

### 2.1.1 Simplex

*Simplex* is a one-way communication such as broadcasting. Figure 2.1 shows a simplified diagram of a simplex communication system. The communication is in one direction only and requires only one communication channel.

© The Author(s), under exclusive license to Springer Nature Switzerland AG. 2019
S. Faruque, *Radio Frequency Multiple Access Techniques Made Easy*,
SpringerBriefs in Electrical and Computer Engineering,
https://doi.org/10.1007/978-3-319-91651-4_2

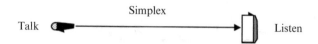

**Fig. 2.1** Simplex communication. The communication is in one direction only and requires only one communication channel

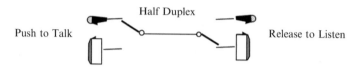

**Fig. 2.2** Half-duplex communication. Push to talk and release to listen is the mechanism of communication. The communication is in two directions, one at a time. It uses one communication channel. It is bandwidth efficient

**Fig. 2.3** Full-duplex
communication. The
communication is in two
directions simultaneously. It
uses two communication
channels

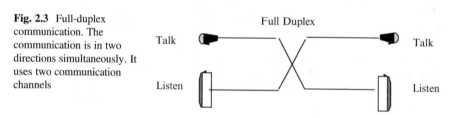

## 2.1.2 Half-Duplex

*Half-duplex* is a two-way communication but one at a time. Push to talk and release to listen is the mechanisms of half-duplex communication systems such as taxi dispatch, walkie-talkie, etc. Figure 2.2 shows a conceptual diagram of a half-duplex communication system. As shown in the figure, the communication is in two directions, one at a time. It uses one communication channel. It is bandwidth efficient.

## 2.1.3 Full Duplex

*Full duplex* is a method of two-way communication, where two users can communicate with each other simultaneously such as land-mobile telephone system. The communication is maintained in both directions at the same time, requiring two channels. The scheme is presented in Fig. 2.3.

In full-duplex communication, the link is maintained in both directions, either in the frequency domain or in the time domain. This is governed by two communication schemes: frequency division duplex (FDD) and time division duplex (TDD). A brief description of these communication schemes is presented in this chapter.

## 2.2   FDD and TDD Schemes

### 2.2.1   Background

Cellular communication is a full-duplex, multiple access communication system (see Fig. 2.4). According to the communication protocol, the cellular base station assigns a carrier frequency to the cell phone. Once the carrier frequency is assigned, the cell phone modulates the carrier frequency by means of voice, data, or video. It then amplifies the modulated carrier frequency and sends it to the antenna for transmission.

Upon receiving, the cellular radio at the base station responds to the cell phone by means of a similar communication protocol. The link is maintained in both directions, either in the frequency domain or in the time domain. This is governed by:

- Frequency division duplex (FDD)
- Time division duplex (TDD)

The two schemes are both widely used. Some cellular systems use TDD, while others use FDD. Some standards also allow for the use of either as both FDD and TDD have their own advantages and disadvantages. A brief description of these communication modes are presented below:

### 2.2.2   Frequency Division Duplex (FDD)

*In FDD*, all the available channels are divided into two bands (lower band and upper band) and grouped as pairs (L1U1, L2U2... LnUn), as depicted in Fig. 1.3 [3, 4]. As can be seen, FDD uses two different frequencies, one for the upload and the other for the download, separated by a guard band. As a result, both transmissions can take place at the same time without interference. This scheme is known as FDMA-FDD technique (Fig. 2.5). A brief description of FDD communication is presented below:

**Fig. 2.4** Illustration of full-duplex cellular communication

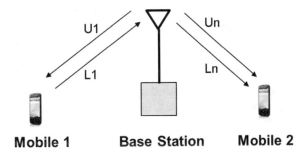

**Mobile 1**      **Base Station**      **Mobile 2**

**Fig. 2.5** Frequency division duplex (FDD). Each user is assigned a pair of frequencies

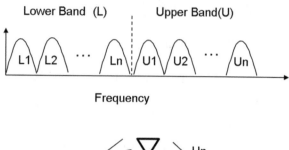

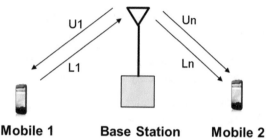

- The base station modulates the carrier frequency (U1) from the upper band and sends the modulated carrier to the mobile. The input modulating signal can be either analog or digital.
- Since the mobile is tuned to the same carrier frequency, it receives the modulated carrier from the base after a propagation delay. It then demodulates the carrier and recovers the information signal.
- In response, the mobile modulates a different carrier frequency (L1) from the lower band and transmits back to the base.
- The base station receives the modulated signal from the mobile, demodulates, and recovers the information.
- The process continues until one of the transmitters terminates the call.

### 2.2.3  Time Division Duplex (TDD)

*In TDD*, a single frequency is time shared between the uplink and the downlink. The duration of transmission in each direction is generally short, in the order of ms. In this scheme, when the mobile transmits, the base station listens, and when the base station transmits, the mobile listens. This is accomplished by formatting the data into a "frame," where the frame is a collection of several time slots. Each time slot is a package of data, representing digitized voice, digitized text, digitized video, and synchronization bits (sync. bits). The sync bits are unique, which is used for frame synchronization. This scheme is known as FDMA-FDD technique. Details will be presented in subsequent *chapters*. Figure 2.6 demonstrates a typical TDD scheme.

**Fig. 2.6** TDD Scheme. A single frequency is time shared between uplink and downlink. This is governed by a frame structure. (**a**) Frame structure. (**b**) Frame transmission scheme

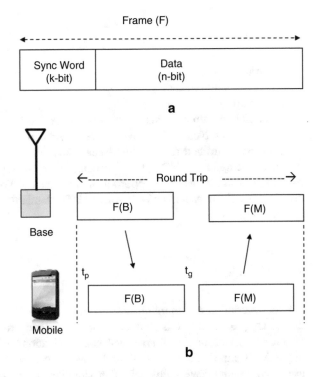

Figure 2.6 illustrates a typical frame and the TDD transmission scheme. According to TDD transmission protocol, both the base station and the mobile use the same carrier frequency. The transmit/receive mechanism between the base station and mobile is as follows:

- The base station modulates the carrier frequency by means of the digital information bits in frame F(B) and transmits to the mobile via the downlink.
- Since the mobile is tuned to the same carrier frequency, it receives the frame F (B) after a propagation delay $t_p$.
- Mobile synchronizes the frame using the sync bits and downloads the data.
- After a guard time $t_g$, the mobile transmits its own frame F(M) to the base via the uplink.
- Base receives the frame F(M) from the mobile after a propagation delay tp an, maintains sync using the sync bi5ts, and downloads the da5ta.
- A round-trip communication is now complete.
- The communication continues until one terminates the call.

As can be seen, the TDD schemes require a propagation delay and a guard time between transmission and reception. The complete round-trip delay $T_d$ must be sufficient to accommodate the frame, propagation delay, and the guard time. Therefore, the round-trip delay can be written as

$$T_d = 2\big(F + t_p + t_g\big) \tag{2.1}$$

where:

$F$ = frame length
$t_p$ = propagation delay
$t_g$ = guard time

In cellular communications, such as OFDMA and LTE (4G), the traffic in both directions is not balanced [5, 6]. There is scheduling protocol, which can be dynamically controlled to offer high-speed data over the downlink and low-speed data over the uplink. In TDD, this is accomplished by transmitting more time slots over the downlink, thereby supporting more capacity. For these reasons, TDD is used in 4G cellular system as WiMAX and LTE standards. Details will be presented in later chapters.

## 2.3   Conclusions

A brief description of communication modes such as simplex, half-duplex, and full duplex are presented in this chapter. Simplex is a one-way communication such as broadcasting. Half-duplex is a two-way communication but one at a time such as taxi dispatch system. Full duplex is referred to as a two-way communication, where two users can communicate with each other simultaneously. The link is maintained in both directions, either in the frequency domain or in the time domain. This is governed by two communication schemes: frequency division duplex (FDD) and time-division duplex (TDD). These communication schemes as applied in the 4G cellular system are also described in this chapter.

## References

1. Milnor, J. W., & Randall, G. A. (2000). Simplex. In *The IEEE authoritative dictionary of standard terms* (7th ed., p. 1053). US: Institute of Electrical and Electronic Engineers.
2. Milnor, J. W., & Randall, G. A. (1931, May). *The Newfoundland-Azores high-speed duplex cable*. Canada: A.I.E.E. Electrical Engineering.
3. IS-54. (1989, December). *Dual-mode mobile station-base station compatibility standard* (EIA/TIA project number 2215).
4. IS-95. (1993, March 15). *Mobile station - base station compatibility standard for dual mode wide band spread spectrum cellular systems* (TR 45, PN-3115).
5. ITU-R. *Requirements related to technical performance for IMT-Advanced radio interface(s)* (Report M.2134), Approved in November 2008.
6. Parkvall, S., & Astely, D. (2009, April). The evolution of LTE toward LTE advanced. *Journal of Communications, 4*(3), 146–154. https://doi.org/10.4304/jcm.4.3.146-154.

# Chapter 3
# Frequency Division Multiple Access (FDMA)

**Abstract** FDMA (frequency division multiple access) is the oldest communication technique used in broadcasting, land-mobile two-way radio, etc. It begins with a band of frequencies, which is allocated by the FCC (Federal Communications Commission). This band of frequency is further divided into several narrow bands of frequencies, where each frequency, also known as channel, is used for full-duplex communication. The communication link is maintained in both directions, either in the frequency domain or in the time domain. This is governed by two basic modes of operations known as the Frequency Division Duplex (FDD) and Time Division Duplex (TDD). These topics, along with FDMA spectrum management and its attributes, are presented in this chapter.

## 3.1 Introduction to FDMA

FDMA (frequency division multiple access) is the oldest communication technique used in broadcasting, land-mobile two-way radio, etc. [1–5, 8, 9]. It begins with a band of frequencies, which is allocated by the FCC (Federal Communications Commission). FCC provides licenses to operate wireless communication systems over the given bands of frequencies. These bands of frequencies are further divided into several channels and assigned to users for full-duplex communication. Figure 3.1 illustrates the basic principle of a typical FDMA technique.

As shown in the figure, the FCC-allocated frequency band is divided into several frequencies, also known as channels. Each channel is assigned to a single user. In this scheme, the channel is occupied for the entire duration of the call. The communication link is maintained in both directions, either in the frequency domain or in the time domain. This is governed by two basic mode of operations listed below:

- Frequency Division Duplex (FDD)
- Time Division Duplex (TDD)

*In FDD*, all the available channels are divided into two bands, lower band and upper band, and grouped as pairs, one for the upload and the other for the download,

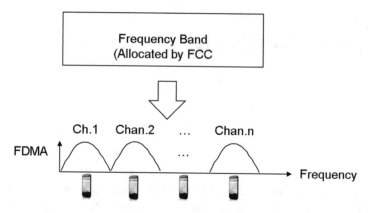

**Fig. 3.1** Illustration of a typical FDMA technique. The FCC-allocated frequency band is divided into several channels. Each channel is assigned to a pair of user

separated by a guard band. As a result, both transmissions can take place at the same time without interference. This scheme is known as FDMA-FDD technique.

*In TDD*, a single frequency is time shared between the uplink and the downlink. In this scheme, when the mobile transmits, base station listens and when the base station transmits, mobile listens. This is accomplished by formatting the data into a "Frame," where the frame is a collection of several time slots. Each time slot is a package of data, representing digitized voice, digitized text, digitized video, and synchronization bits (sync. bits). The sync bits are unique, which are used for frame synchronization. This scheme is known as FDMA-FDD technique.

This chapter will present a brief overview of FDMA-FDD and FDMA-TDD techniques used in the cellular system. The concept of frequency reuse will be studied along with cell reuse plan [3] related to OMNI and sectorization schemes [5].

## 3.2  FDMA-FDD Technique

In FDMA-FDD, all the available channels are divided into two bands, lower band and upper band, and grouped as pairs as shown below: L1U1, L2U2, ..., Ln Un. This is shown in Fig. 3.2. As can be seen in the figure, FDD uses two different frequencies, one for the upload and the other for the download, separated by a guard band. As a result, both transmissions can take place at the same time without interference. This scheme is known as FDMA-FDD technique.

A brief description of FDMA-FDD communication, as implemented in 1G cellular system [1–3], is presented below:

- The base station modulates the carrier frequency (U1) from the upper band and sends the modulated carrier to the mobile. The input modulating signal can be either analog or digital.

**Fig. 3.2** Frequency
Division Duplex (FDD)
technique. Both frequencies
can operate at the same time
without interference

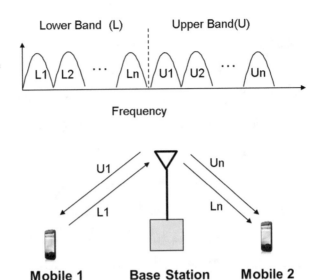

- Since, the mobile is tuned to the same carrier frequency, it receives the modulated carrier from the base after a propagation delay. It then demodulates the carrier and recovers the information signal.
- In response, the mobile modulates a different carrier frequency (L1) from the lower band and transmits back to the base.
- The base station receives the modulated signal from the mobile, demodulates and recovers the information.
- The process continues until one of the transmitters terminates the call.

Figure 3.3 shows the conceptual realization of an FDMA radio operating in the FDD mode. In this mode of operation, a pair of carrier frequencies (e.g., L1 U1) is used at the same time during a call, one from the lower band and one from the upper band. The lower band frequency is used by the mobile radio, and the upper band frequency is used by the base station radio. It may be noted that, lower frequencies are used in the mobile because propagation decay is logarithmic as a function of frequency. Since the mobile transmit power is much lower than the base station radio, it is necessary to use lower frequencies for the mobile phone. FDMA is the multiple access technique used in AMPS (Advanced Mobile Phone System), also known as first-generation (1G) cellular communication system.

## 3.3 FDMA-TDD Technique

In FDMA-TDD, a single FDMA frequency is time shared between the uplink and the downlink. The duration of transmission in each direction is generally short, in the order of ms. In this scheme, when the mobile transmits, base station listens, and

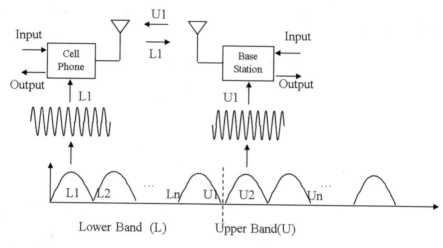

**Fig. 3.3** FDMA-FDD radio. A pair of channels is used during a call: one from the lower band and one from the upper band

when the base station transmits, mobile listens. This is accomplished by formatting the data into a "Frame," where the frame is a collection of several time slots. Each time slot is a package of data, representing digitized voice, digitized text, digitized video, and synchronization bits (sync. bits). The sync bits are unique, which is used for frame synchronization. Figure 3.4 illustrates a typical frame and the TDD transmission scheme.

According to TDD transmission, both the base station and the mobile use the same carrier frequency. The transmit/receive mechanism between the base station and mobile is as follows:

- The base station modulates the carrier frequency by means of the digital information bits in frame F(B) and transmits to the mobile.
- Since the mobile is tuned to the same carrier frequency, it receives the frame F (B) after a propagation delay $t_p$.
- Mobile synchronizes the frame using the sync bits and downloads the data.
- After a guard time $t_g$, mobile transmits its own frame F(M) to the base using the same carrier frequency.
- Base receives the frame from the mobile after a propagation delay $t_p$, maintains sync using the sync bits, and downloads the data.
- A round-trip communication is now complete.
- The communication continues until one terminates the call.

As can be seen in the figure, the TDD schemes require a propagation delay and a guard time between transmission and reception. The complete round-trip delay $T_d$ must be sufficient to accommodate the frame, the propagation delay, and the guard time. Therefore, the round-trip delay can be written as,

**Fig. 3.4** TDD frame structure. (**a**) Frame. (**b**) Frame transmission scheme

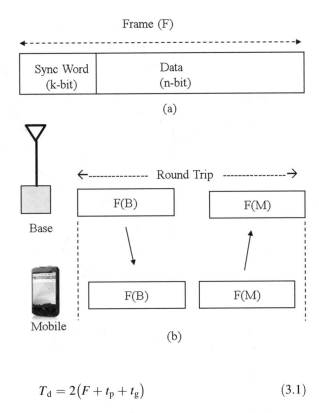

(a)

(b)

$$T_d = 2(F + t_p + t_g) \qquad (3.1)$$

The round-trip delay $T_d$ depends on the frame length $F$, which is generally in milliseconds (ms). The propagation delay $t_p$ depends on the propagation distance, and the guard time $t_g$ depends on the technology.

Figure 3.5 shows a conceptual realization of an FDMA radio operating in the TDD mode. In this technique, a single FDMA frequency is used in both directions during the entire duration of the call. This is accomplished by means of a special frame structure as depicted in the figure. In this scheme, when the mobile transmits, base station listens, and when the base station transmits, mobile listens. This is bandwidth efficient, since a single-carrier frequency is used during the entire duration of the call.

In cellular communications, such as OFDMA and LTE (4G), the traffic in both directions is not balanced. The volume of data transmission can be dynamically adjusted in each direction by means of FDD technique. There is scheduling protocol, which can be dynamically controlled to offer high-speed data over the downlink and low-speed data over the uplink. In TDD, this is accomplished by transmitting more time slots over the downlink, thereby supporting more capacity. For these reasons, TDD is used in 4G cellular system as WiMAX and LTE standards [6, 7].

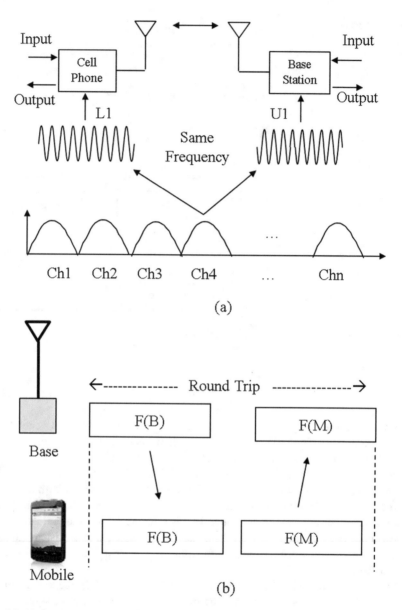

**Fig. 3.5** FDD-TDD Radio. A single FDMA channel is used in both directions

## 3.4   FDMA Cellular Telephoney

Cellular communication is a full-duplex or simply a duplex communication system. Duplex is referred to as two-way communication, where two users can communicate with each other simultaneously. Figure 3.6 illustrates the basic concept of a full-

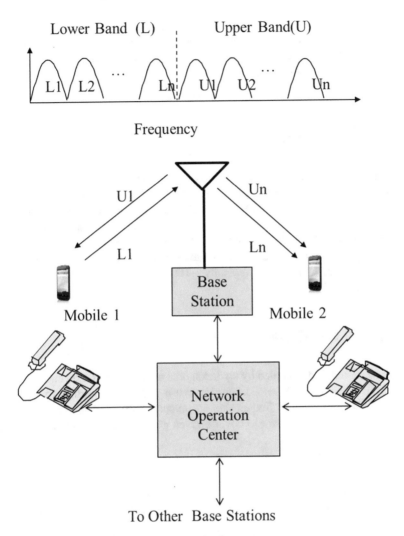

**Fig. 3.6**  The 1G Cellular FDMA-FDD communication system

duplex cellular communication system, developed as the 1st-generation (1G) cellular communication system, based on FDMA-FDD technique [1–5].

Each mobile is assigned a pair of channels, one from the lower band and the other from the upper band, enabling simultaneous transmission and reception. The channel assignment is governed by the network operation center (NOC). The NOC is also connected to other cellular base stations and landline telephone system. As can be seen in the figure, multiple users have access to the same base station. Each user occupies two different frequencies, one for the upload and the other for the download, separated by a guard band. As a result, both transmissions can take place at the same time without interference. For these reasons, this scheme is known as FDMA-

FDD technique. This forms the basis of 1G cellular communication system such as North American AMPS (Advanced Mobile Phone System) [1, 2] and European GSM (Group Special Mobile). This heralded a new era in cellular communication. Let's take a closer look!

## 3.5 The Concept of Cell

A cell is a geographical area covered by radio-frequency (RF) signals. The RF source is located at the center of the cell. Since RF propagation is fuzzy due to irregular terrain, vegetation, hills, buildings, etc., a practical cell will be highly irregular. As a result, it will be difficult to analyze. For these reasons a hexagonal cellular geometry is used for planning and system dimensioning. This is depicted in Fig. 3.7. The signal strength at the cell edge is the minimum signal strength that the mobile can detect. This value is typically $-100$ dBm.

## 3.6 $N = 7$ Cell Cluster

A hexagonal cell is surrounded by six hexagonal cells, forming a seven-cell cluster. This is also known as $N = 7$ frequency reuse pattern. According to frequency planning, all the available frequencies are evenly distributed among seven cells and then reused over and over to cover a given geographical area. This is depicted in Fig. 3.8.

**Fig. 3.7** Illustration of a hexagonal cell, which is used for system dimensioning

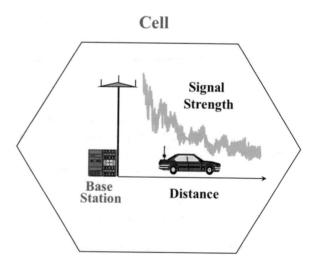

**Fig. 3.8** A 7 cell cluster. All the available frequencies are evenly distributed among seven cells

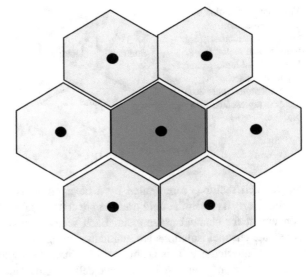

**Fig. 3.9** A three-cell cluster. All the available frequencies are evenly distributed among three cells

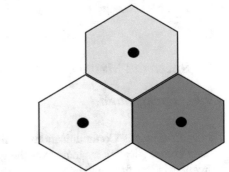

## 3.7  $N = 3$ Cell Cluster

Figure 3.9 shows a cluster of three cells. All the available channels are equally distributed among three cells. This is known as $N = 3$ frequency planning. According to frequency planning, all the available frequencies are evenly distributed among three cells and then reused over and over to cover a given geographical area. Consequently, $N = 3$ frequency planning offers more capacity.

## 3.8  Sectorization

The 120° sectorization is achieved by dividing a cell into three sectors, 120 degrees each, as shown in Fig. 3.10a. Each sector is treated as a logical cell, where directional antennas are used in each sector. Figure 3.10b shows an alternate representation,

**Fig. 3.10** Illustration of 120° sectorization. Directional antennas are used in each sector. (**a**) Conventional representation (**b**) an alternate representation, where each sector is represented by a hexagon

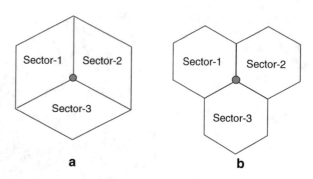

a                                                        b

where each sector is represented by a hexagon. This configuration is known as tri-cellular plan [5]. Both configurations are conceptually identical, while the latter is convenient for channel assignment. Each sector uses a set of different channels. Adequate channel isolations are maintained within and between sectors in order to minimize interference. This is attributed to channel assignment techniques. Because directional antennas are used in sectored cells, it allows antenna down tilt to improve C/I and channel capacity [5].

## 3.9  Spectrum Management

### 3.9.1  Background

The FCC (Federal Communications Commission) provides licenses to operate cellular communication systems over the given bands of frequencies. These bands of frequencies are finite and have to be reused to provide services to other geographic areas. In frequency reuse plans, we dole out the channels to the cells, much like a dealer in card game deals out cards from the deck until every player has a set. Several frequency reuse techniques, generally known as frequency planning or channel assignment techniques, are available. Some of the most widely used frequency planning techniques are given below [5, 10–15]:

- $N = 7$ Frequency Reuse Plan
- $N = 3$ Frequency Reuse Plan

### 3.9.2  The N = 7 Frequency Planning

Figure 3.11 shows a frequency plan in clusters of seven cells, known as $N = 7$ frequency plan. In this plan, all the available channels are evenly distributed among seven cells. This is the first arrangement which works in most propagation environments. There are also other frequency reuse plans, for example, $N = 3, N = 4, N = 9$,

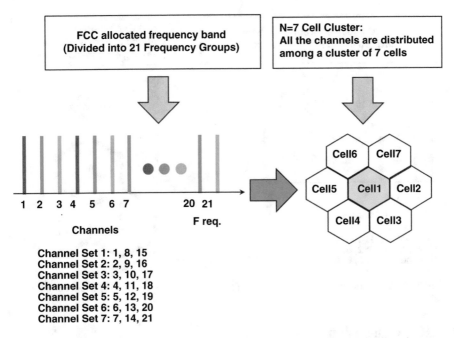

Channel Set 1: 1, 8, 15
Channel Set 2: 2, 9, 16
Channel Set 3: 3, 10, 17
Channel Set 4: 4, 11, 18
Channel Set 5: 5, 12, 19
Channel Set 6: 6, 13, 20
Channel Set 7: 7, 14, 21

**Fig. 3.11** $N = 7$ frequency plan. All the available channels are evenly distributed among seven cells

etc. However, an increase in the number of cells per cluster decreases cell capacity. For this reason, the $N = 3$ is the most widely used frequency plan today.

The $N = 7$ channel assignment is based on the following sequence: ($N$, $N + 7$, $N + 14$) where $N$ is the cell number ($N = 1, 2, 7$). The channel grouping scheme is shown in the inset of Fig. 3.10. Each frequency group has several frequencies (known as channels). These 21 frequency groups are equally distributed among 7 cells, 3 frequency groups per cell. Notice that each OMNI cell gets 3 frequency groups or a total of 21 frequency groups in the 7 cell cluster. Note that OMNI means all directions; the base station antenna is located at the center of the cell.

### 3.9.3  N = 3 Frequency Planning

Figure 3.12 shows a frequency reuse in cluster of three cells, known as $N = 3$ frequency planning. This is the most widely used frequency reuse plan today. In this plan all the available channels are equally distributed among three cells. There are also other frequency reuse plans, for example, $N = 4$, $N = 7$, $N = 9$, etc. However, an increase in the number of cells per cluster decreases cell capacity. The $N = 3$ channel assignment is based on the following sequence: ($N$, $N + 3$, $N + 6$) where $N$ is the cell number ($N = 1, 2, 3$). The channel grouping scheme is shown in the inset of Fig. 3.12. Each frequency group has several frequencies (known as channels).

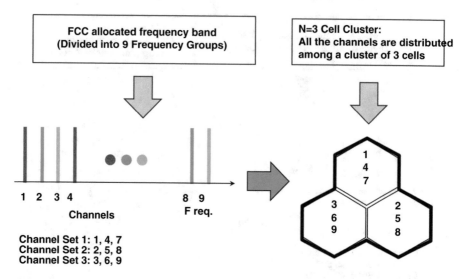

**Fig. 3.12** $N = 3$ frequency plant. All the available channels are evenly distributed among three cells

## 3.10 Conclusions

FDMA (frequency division multiple access) is the oldest communication technique used in broadcasting, land-mobile two-way radio, etc. It begins with a band of frequencies, which is allocated by the FCC (Federal Communications Commission). This band of frequency is further divided into several narrow bands of frequencies, where each frequency, also known as channel, is used for full-duplex communication. The communication link is maintained in both directions, either in the frequency domain or in the time domain. This is governed by two basic modes of operations known as Frequency Division Duplex (FDD) and Time Division Duplex (TDD).

This chapter presents a brief overview of FDMA along with FDD and TDD mode of operations. The FDMA cellular communications is described along with various spectrum management techniques.

## References

1. IS-54. (1989, December). *Dual-mode mobile station-base station compatibility standard* (EIA/TIA project number 2215).
2. IS-95. (1993, March 15). *Mobile station - base station compatibility standard for dual mode wide band spread spectrum cellular systems* (TR 45, PN-3115).
3. Mac Donald, V. H. (1979, January). The cellular concept. *The Bell System Technical Journal*, 58(1), IS–41.
4. Lee, W. C. Y. (1989). *Mobile cellular telecommunications systems.* New York: McGraw-Hill.

5. Faruque, S. (1996). *Cellular mobile systems engineering*. Norwood: Artech House, ISBN: 0-89006-518-7.
6. ITU-R. *Requirements related to technical performance for IMT-Advanced radio interface(s)* (Report M.2134), Approved in November 2008.
7. Parkvall, S., & Astely, D. (April 2009). The evolution of LTE toward LTE advanced. *Journal of Communications, 4*(3), 146–154. https://doi.org/10.4304/jcm.4.3.146-154.
8. ITU, International Telecommunications Union, Paris, France.
9. FCC, Federal Communication Commission, Washington, DC.
10. Faruque, S. *Directional frequency allocation in an N=6 cellular radio system* (US Patent: 5,802,474), Granted, 1 September 1998.
11. Faruque, S. *Frequency assignment in a cellular radio system* (US Patent: 5,734,983), Granted 31 March 1998.
12. Faruque, S. *Frequency plan for a cellular network* (US Patent: 5,483,667), Granted, 9 January 1996.
13. Faruque, S. *High capacity cell planning based on fractional frequency reuse* (US Patent: 6,128,497), Granted, 3 October 2000.
14. Faruque, S. *Frequency transition process for capacity enhancement in a cellular network* (US Patent: 6,085,093), Granted, 4 July 2000.
15. Faruque, S. *N=4 directional frequency assignment in a cellular radio system* (US Patent: 5,970,411), Granted, 19 October 1999.

# Chapter 4
# Time Division Multiple Access (TDMA)

**Abstract** Time division multiple access (TDMA) is a method of transmitting and receiving multiple independent signals over a single transmission channel. The TDMA at the transmit side, known as the multiplexer, assigns multiple channels in preassigned time slots. The TDMA at the receive side, known as the de-multiplexer, separates the incoming composite signal into parallel streams. Both multiplexer and de-multiplexer are synchronized by a common clock to receive data in accordance with the transmit sequence. This chapter presents the key concepts, underlying principles and practical applications of TDMA used in land-mobile telecommunication systems.

## 4.1 Introduction to TDMA

TDMA (time division multiple access) for wireless communication is an extension of FDMA, where each FDMA channel is time shared by multiple users, one at a time. It begins with a band of frequencies, which is allocated by the FCC (Federal Communications Commission). This band of frequency is further divided into several narrow bands of frequencies, where each frequency, also known as channel, is used for full-duplex communication by multiple users one at a time as depicted in Fig. 4.1.

Figure 4.2 illustrates the basic concept of a full-duplex cellular communication system, developed as the 2nd-generation (2G) cellular communication system, based on TDMA-FDD technique [1–6]. In this technique, a pair of FDMA channel is used during a call, one from the lower band and one from the upper band. The lower band frequency is time shared by several mobiles. The upper band frequency is also time shared synchronously by the base station radio. Both channels are occupied during the entire duration of the call.

Synchronization is achieved by means of a special frame structure, where the frame is a collection of time slots. Each time slot is assigned to a mobile. This implies that, when one mobile has access to the channel, the other mobiles are idle.

© The Author(s), under exclusive license to Springer Nature Switzerland AG. 2019    35
S. Faruque, *Radio Frequency Multiple Access Techniques Made Easy*,
SpringerBriefs in Electrical and Computer Engineering,
https://doi.org/10.1007/978-3-319-91651-4_4

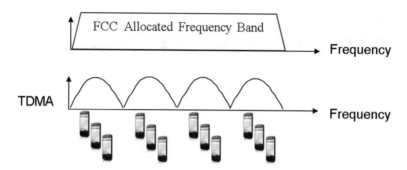

**Fig. 4.1** North American Cellular TDMA technique. The FCC-allocated frequency band is divided into several channels where each channel is time shared by several users one at a time

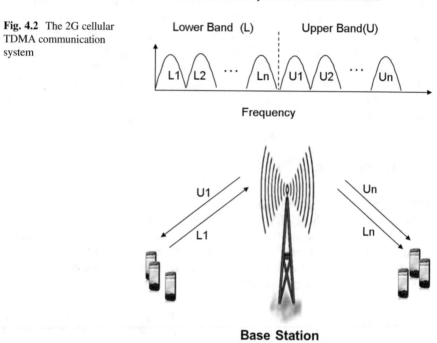

**Fig. 4.2** The 2G cellular TDMA communication system

Therefore, TDMA synchronization is critical for data recovery and collision avoidance.

TDMA has several advantages over FDMA:

- Increased channel capacity
- Greater immunity to noise and interference
- Secure communication
- More flexibility and control

Moreover, it allows the existing FDMA standard to coexist in the same TDMA platform, sharing the same RF spectrum.

## 4.2  TDMA Frame Structure

The TDMA air link is based on a 40 ms frame structure, equally divided into six time slots, 6.667 ms each. Each of the six time slots contains 324 gross bit intervals, corresponding to 162 symbols (1 symbol = 2 bits of information). Figure 4.3 shows the forward link (base to mobile) TDMA frame structure. In TDMA-3, the time slots are paired as 1–4, 2–5, and 3–6 where each disjoints pair of time slots are assigned to a mobile. This arrangement enables three mobiles to access the same 30 kHz channel one at a time.

The TDMA-3 forward link uses a rate 1/2 convolutional encoding with interleaving. The encoded 48.6 kb/s data bit stream is modulated by means of a $\pi/4$ DQPSK modulation and then transmitted from the base station to the mobile where each mobile receives data at 16.2 kb/s. At the receive side the RF signal is demodulated and decoded, and finally the original data is recovered. Since this is a radio channel, the recovered data is impaired by noise, interference, and fading. As a result, the information is subject to degradation. Although error control coding greatly enhances the performance, the C/I (carrier-to-interference ratio) is still the limiting factor. The TDMA-3 reverse link is exactly the reverse process.

**Problem 4.1**
Given:
- Frame length = 40 ms (Figure below).
- The frame contains six time slots and supports three users.
- Each user originates 16.2 kbps data.

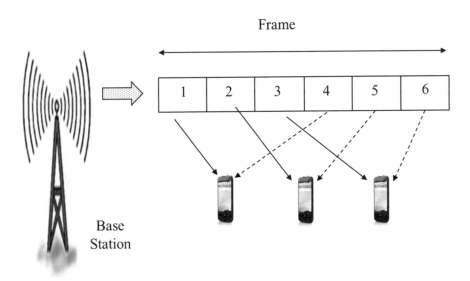

**Fig. 4.3** TDMA forward link format

Frame =40 ms

| 1 | 2 | 3 | 4 | 5 | 6 |
|---|---|---|---|---|---|

Find:

(a) A suitable multiplexing structure
(b) The composite data rate in the channel
(c) Number of bits/Frame

Solution:

(a) We have three users and six time slots. Therefore we can assign two time slots/user:

    • User-1: Time slot 1 and 4
    • User-2: Time slot 2 and 5
    • User-3: Time slot 3 and 6

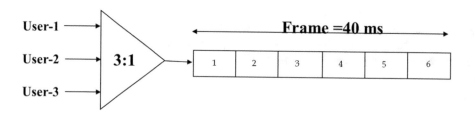

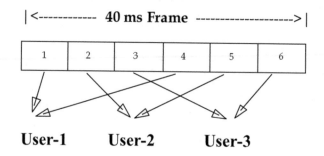

User-1      User-2      User-3

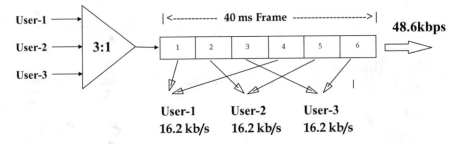

(b)  Composite Data Rate:

$$16.2 \ \text{kbps/user} \times 3 = 48.6 \ \text{kbps}$$

(c)  Number of bits/Frame = Frame length/Bit duration

$$= 40 \ \text{ms}/(1/48.6 \ \text{kbps}) = 1944 \ \text{bits/Frame}$$

## 4.3   European GSM TDMA Cellular

### 4.3.1   GSM TDMA Schème

GSM (Groupe Spécial Mobile), also known as Global System for Mobile Communications, is a second-generation (2G) digital cellular communication standard developed in Europe [7–10]. In GSM, a given frequency band is divided into 200 KHz per carrier, where each carrier is time shared by eight users as depicted in Fig. 4.4.

Since each RF carrier frequency is time shared between eight users, there are eight time slots in the GSM frame. According to the GSM communication protocol, the time slot is assigned to a mobile during the channel assignment session. Therefore, when one mobile is active, other mobiles remain idle. Each mobile waits for their turn.

### 4.3.2   GSM TDMA Frame (4.615 ms)

The GSM frame is constructed by multiplexing eight time slots as shown in Fig. 4.5. It has the following design specifications:

- Frame duration: 4.615 ms
- Frame length: eight time slots

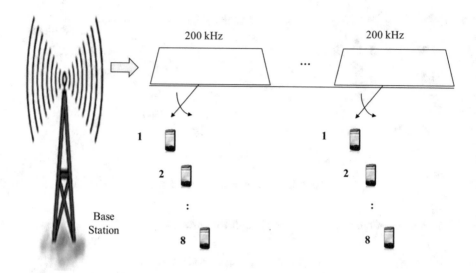

**Fig. 4.4** The European GSMTDMA scheme. Each 200 kHz channel is time shared by eight users

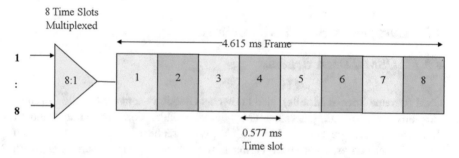

**Fig. 4.5** GSM TDMA frame structure. Eight time slots multiplexed to construct a frame

- Duration of time slot: 0.576875 ms = 0.577 ms
- Number of bits per time slot: 156.25 bits
- Bit rate: 270.833 kb/s

### 4.3.3  GSM Multi-frame

The GSM TDMA hierarchy is composed of two types of multi-frame structure known as:

- Control multi-frame for messaging and signaling (51 time slots, 0.577 ms/time slot)
- Traffic multi-frame for voice and data (26 time slots, 0.577 ms/time slot)

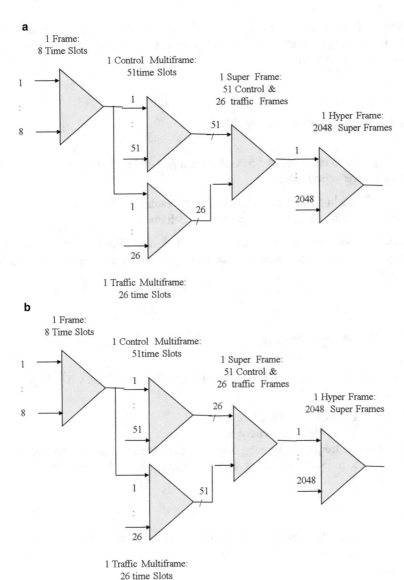

**Fig. 4.6** GSM frame hierarchy composed of frame, multi-frame, superframe, and hyperframe

The control multi-frame, shown in Fig. 4.6, has 51 time slots – called control multi-frame, composed of 51 bursts in duration of 234.4 ms. The time slots are used for messaging and controlling function such as channel assignment, handoff, paging, etc.

The traffic multi-frame, also shown in Fig. 4.6, has 26 time slots – called traffic multi-frame, composed of 26 bursts in duration of 120 ms. These time slots are used voice and data communication.

### 4.3.4   GSM Superframe (6.12 s)

GSM superframe is constructed by multiplexing several multi-frames. There are two schemes in the superframe:

- 51 traffic multi-frames and 26 control multi-frames
- 26 traffic multi-frames and 51 control multi-frames

When the scheme interchanges, the different number of traffic and control multi-frames within the superframe, the time interval within the superframe remains the same.

### 4.3.5   GSM Hyperframe

GSM hyperframe is constructed by multiplexing 2048 superframes. The hyperframe is used to:

- Maintain different schedules in order.
- Synchronize and maintain encryption.
- Synchronize frequency hopping between the transmitter and the receiver ao that they hop to a new frequency at the same time. This is a low frequency hopping feature in GSM.

**Problem 4.2**

Given:

A 64 kb/s serial data is converted into eight parallel streams. What is the output bit rate after serial to parallel conversion?

Solution:

The output bit rate after 1:8 serial to parallel conversion will be 64 kbps/8 = 8 kb/s as shown in the figure below:

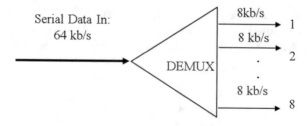

**Problem 4.3**

Given:

Eight parallel data streams, having 8 kb/s each, is parallel to serial converted to form a composite bit stream. What is the output bit rate after 8:1 parallel to serial conversion?

Solution:

The output bit rate after 8:1 parallel to serial conversion will be 8 kb/s × 8 = 64 kb/s as shown in the figure below:

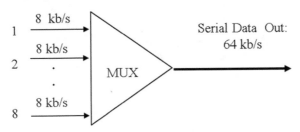

## 4.4   Conclusions

- Defined TDMA
- Described TDMA frame and frame hierarchy used in North American and European Land Telephone System
- Illustrated Frame Synchronization Process
- Reviewed North American TDMA and European GSM TDMA System used in Wireless Communications
- Illustrated TDMA Frame and Frame Hierarchy

## References

1. Smith, D. R. (1984). *Digital transmission systems*. New York: Van Nostrand Reinhold Company, ISBN: 0442009178.
2. Couch, L. W., II. (2001). *Digital and analog communication systems* (7th ed.). Englewood Cliffs, NJ: Prentice-Hall, ISBN: 0-13-142492-0.
3. Faruque, S., & Baxter, S. (1989). *Nortel internal training course, 1995 and IS-54* (EIA Project Number 2215), pp. 3/18–3/47.
4. Mac Donald, V. H. (1979, January) Advanced mobile phone services. Special Issue, *Bell System Technical Journal, 58*, 12–41.
5. Lee, W. C. Y. (1989). *Mobile cellular telecommunications systems*. New York: McGraw-Hill.
6. Faruque, S. (1996). *Cellular Mobile Systems Engineering*. Boston: Artech House, ISBN: 0-89006-518-7.
7. Proakis, J. G., & Salehi, M. (2014). *Fundamenttals of Communication System* (2nd ed.). Boston: Pearson, ISBN-13: 978-0-13-335485-0.
8. European Telecommunications Standards Institute. (2011). GSM. etsi.org.
9. GSM Association. (2001). History. gsmworld.com.
10. European Telecommunications Standards Institute. (2011). Cellular history. etsi.org. Archived from the original on 5 May 2011.

# Chapter 5
# Code Division Multiple Access (CDMA)

**Abstract** This chapter presents a brief overview of spread spectrum technique and shows how it relates to CDMA technology. Orthogonal codes and their properties are presented and show how orthogonal codes are generated and used to design CDMA radio. It is shown that CDMA capacity directly relates to code length.

**Objectives**

- Review spread spectrum and show how it relates to process gain and CDMA capacity.
- Understand orthogonal codes and show how orthogonal codes are used to design CDMA Radio.
- Estimate CDMA capacity and show how it relates to code length.
- Recognize the need for power control and show how it relates to CDMA operation.

## 5.1   Introduction to CDMA

CDMA (code division multiple access) is a spread spectrum (SS) communication system where multiple users have access to the same career frequency at the same time [1–5]. It begins with a frequency band, allocated by the Federal Communication Commission (FCC) as shown in Fig. 5.1. FCC provides licenses to operate wireless communication systems over given bands of frequencies. These frequency bands are finite and have to be reused to support a large number of users in a given geographical area.

Figure 5.2 shows the construction of a simple CDMA radio. At the transmit side, the input encoded low-speed NRZ data (D) is multiplied by a high-speed orthogonal code (C) for spectrum spreading. Multiplication in this process is referred to as exclusive OR (EXOR) operation. The output of the exclusive OR gate Y is then modulated and transmitted through the radio-frequency (RF) channel.

© The Author(s), under exclusive license to Springer Nature Switzerland AG. 2019     45
S. Faruque, *Radio Frequency Multiple Access Techniques Made Easy*,
SpringerBriefs in Electrical and Computer Engineering,
https://doi.org/10.1007/978-3-319-91651-4_5

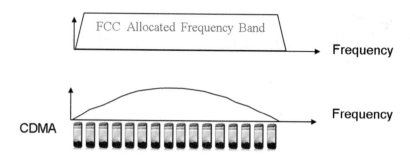

**Fig. 5.1** FCC-allocated frequency band and its use in CDMA as spread spectrum. Multiple users have access to the same frequency at the same time

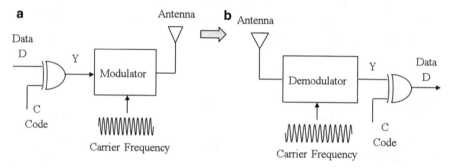

**Fig. 5.2** The basic CDMA radio. (**a**) Transmitter. (**b**) Receiver

At the receive side, the incoming RF signal is demodulated to recover the composite digital signal Y that contains both D and C. The receiver then multiplies Y by using the same orthogonal code C to recover the original data D. The operation is governed by the following Boolean expressions:

- Y = D EXOR C
- D = Y EXOR C
  - = D EXOR C EXOR C
  - = D

where

- D = Input NRZ Data
- C = $n$-bit Orthogonal Code
- EXOR C EXOR C = 0

The objective of this chapter is to review spectrum, spectrum spreading, and dispreading techniques and shows how it relates to spread spectrum CDMA radio. Orthogonal codes and their properties are presented and show how orthogonal codes are used in CDMA radio and to estimate channel capacity. CDMA is a noise-limited system. Therefore, the need for power control is also described for reliable operation.

## 5.2   Spectrum of NRZ Data

Let's consider the discrete time signal as shown in Fig. 5.3, having the following boundary conditions [5]:

$$V(t) = V \quad < 0 < t < T$$
$$= 0 \quad \text{elsewhere} \tag{5.1}$$

This signal is also known as NRZ (non-return-to-zero) data, generally used in digital radio. Now, our goal is to determine the frequency content of this signal and then to evaluate the power spectrum associated with this signal. According to Fourier transform, the time domain signal is given by,

$$V(\omega) = \int_0^T V.e^{-j\omega\,t} dt$$
$$= \left( 2\frac{V}{\omega} \right) \sin(\omega T/2) \tag{5.2}$$
$$= VT \left[ \frac{\sin(\omega T/2)}{\omega T/2} \right]$$

The corresponding power spectral density is given by,

$$P(\omega) = \left( \frac{1}{T} \right) |V(\omega)|^2$$
$$= V^2 T \left[ \frac{\sin(\omega T/2)}{\omega T/2} \right]^2 \tag{5.3}$$

Figure 5.4 is the familiar power spectrum, showing the main lobe corresponding to the fundamental component of the frequency and infinite number of side lobes corresponding to the harmonic components. We also note that most of the power is

**Fig. 5.3** A non-return-to-zero (NRZ) data

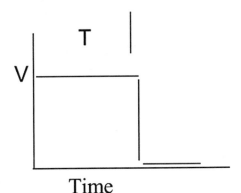

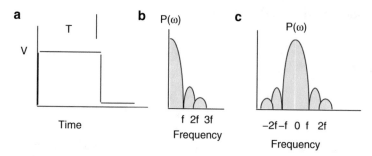

**Fig. 5.4** Power spectrum of NRZ data. (**a**) NRZ data. (**b**) One-sided power spectrum. (**c**) Two-sided power spectrum. The two-sided power spectrum will be shifted and centered on the carrier frequency after modulation

retained by the main lobe whose bandwidth is given by, $BW = 2f$, where $f =$ bit rate. It may be noted that the two-sided power spectrum will be shifted and centered on the carrier frequency after modulation.

## 5.3  Spectrum Spreading

### 5.3.1  Basic Concept

Spectrum spreading can be achieved by simply increasing the frequency of the discrete time signal. Thus we consider a waveform with an amplitude V and frequency f ($f = 1/T$) and then increase the frequency of the same waveform by a factor of $n$, i.e., $T$ is now reduced by $n$. Pair of boundary conditions describing these signals are as follows:

$$
\begin{aligned}
V(t) &= V \quad &< 0 < t < T \\
&= 0 \quad &\text{elsewhere} \\
V(t) &= V \quad &< 0 < t < T/n \\
&= 0 \quad &\text{elsewhere}
\end{aligned}
$$
(5.4)

Applying the Fourier transform in the above equations, we obtain the corresponding frequency domain signals as follows:
For $0 \leq t \leq T$:

$$
V(\omega) = \int_0^T V.e^{-j\omega t}dt = VT\left[\frac{\sin (\omega T/2)}{\omega T/2}\right]
$$
(5.5)

**Fig. 5.5** Illustration of spectrum of two signals having different frequencies

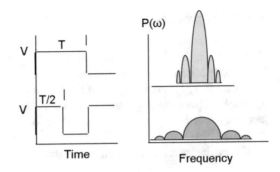

$$P(\omega) = \left(\frac{1}{T}\right)|S(\omega)|^2 = V^2T\left[\frac{\sin(\omega T/2)}{\omega T/2}\right]^2 \tag{5.6}$$

For $0 \leq t \leq T/n$:

$$V(\omega) = \int_0^T V.e^{-j\omega t}dt = VT\left[\frac{\sin(\omega T/2n)}{\omega T/2n}\right]$$

$$\tag{5.7}$$

$$P(\omega) = \left(\frac{1}{T}\right)|S(\omega)|^2 = V^2T\left[\frac{\sin(\omega T/2n)}{\omega T/2n}\right]^2$$

Figure 5.5 shows the power spectrum for $n = 1$ and $n = 2$, where $n1$ relates to the low frequency signal and $n2$ relates to the high frequency signal. In this particular example, the latter is twice the frequency of the former.

## 5.3.2 Energy Delivered

The energy delivered into a load between time $t = 0$ and $t = T$ is given by the total area under the curve.

$$E(t) = \int_0^T P(\omega)dt = \left(\frac{1}{T}\right)\int_0^T |S(\omega)|^2 dt = \text{constant} \tag{5.8}$$

This means that the total energy under the power spectrum curve remains the same after spreading. It follows that if we increase the bandwidth by a certain factor, the amplitude will be reduced by the same factor. This is because of conservation of energy.

### 5.3.3   Process Gain

Process gain is due to spectrum spreading and dispreading. As shown in Fig. 5.6, spectrum spreading occurs at the transmit side by multiplying the input NRZ data (D) of bit rate $R_{b1}$ by an $n$-bit orthogonal code (c) of code rate $R_c$, where $R_c = nR_b$. Spectrum dispreading is done at the receiving side by multiplying the incoming data by the same orthogonal code (C) to recover the original data (D) (Fig. 5.7).

The mathematical expression that describes the process gain is given by:

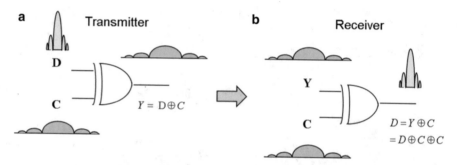

**Fig. 5.6** Illustration of spectrum spreading and dispreading. (**a**) Spectrum spreading at the transmitter. (**b**) Spectrum dispreading at the receiver. The total power is the same before and after spreading due to conservation of energy

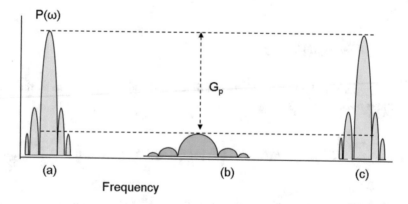

**Fig. 5.7** Illustration of process gain. (**a**) Power spectrum before spreading. (**b**) Power spectrum after spreading and (**c**) Power spectrum after dispreading. Dispreading regenerates the original power spectrum and recovers the NRZ data. The difference in amplitude between the spread and the un-spread spectrum is the process gain $G_p$

**Table 5.1** dB margins due to spectrum spreading

| Code length ($n$) | Process gain (dB) |
|---|---|
| 2 | 3.0103 |
| 4 | 6.0206 |
| 8 | 9.0309 |
| 16 | 12.0412 |
| 32 | 15.0515 |
| 64 | 18.0618 |

$$G_p(dB) = 10\log\left(\frac{R_c}{R_b}\right)$$
$$= 10\log(n)$$

(5.9)

where

- $G_p$ = Process gain
- $R_c$ = Code rate
- $R_b$ = Bit rate
- $n$ = Code length

For example, if the orthogonal code length is 2 ($n = 2$), the corresponding process gain will be $G_p(dB) = 10\log(2) = 3$ dB. This is an additional dB margin, which improves the signal to noise ratio (S/N) by 3 dB. Similarly, if the orthogonal code length is 64 ($n = 64$), the corresponding process gain will be $G_p(dB) = 10\log(64) = 18$ dB, and the additional margin in S/N is 18 dB. Table 5.1 shows the dB margins associated with a few orthogonal codes.

It appears that, in CDMA, process gain improves S/N, which leads to CDMA capacity.

## 5.4  A Tour of Orthogonal Codes

Orthogonal codes, also known as Walsh codes, were originally developed by J. L. Walsh in 1923. Walsh codes are well known for their orthogonal properties. They have been successfully implemented in CDMA for spreading and user ID. The use of orthogonal codes for Forward Error Control Coding (FECC) has also been investigated by a limited number of authors. In this chapter, our goal is to show how orthogonal codes are generated and used in CDMA to support multiple users at the same time in a given channel.

## 5.4.1   Generation of Orthogonal Codes

Orthogonal codes are binary valued and can be generated by means of an $N \times N$ Hadamard matrix as follows [5]:

- Construct an $N \times N$ matrix as four quadrants

<u>N x N Matrix</u>

| 1st Quadrant | 2nd Quadrant |
|---|---|
| 3rd Quadrant | 4th Quadrant |

- Keep the 1st, the 2nd, and the 3rd quadrants identical and invert the 4th as follows:

| b | b |
|---|---|
| b | $\overline{b}$ |

Where $b$ is a binary bit which can be either 0 or 1. This process governs the generation of an $N \times N$ Hadamard matrix for $N$-orthogonal codes with $b = 0$ and an $N \times N$ Hadamard matrix for an $N$-bit antipodal code with $b = 1$.

For example, a $2 \times 2$ Hadamard matrix generates two orthogonal codes and two antipodal codes, for a total of four bi-orthogonal codes as follows:

Hadamard   Orthogonal   Antipodal
  Matrix     Code block   Code block

| b | b |   | 0 | 0 |   | 1 | 1 |
|---|---|   |---|---|   |---|---|
| b | $\underline{b}$ |   | 0 | 1 |   | 1 | 0 |

Where, a 2-× 2 Hadamard matrix generates two orthogonal codes, having two bits each as shown below:

| Two-bit orthogonal code block | Two-bit antipodal code block |
|---|---|
| 0 0 | 1 1 |
| 0 1 | 1 0 |

Similarly, a 4 × 4 Hadamard matrix generates four orthogonal codes and four antipodal codes, for a total of eight bi-orthogonal codes as follows:

Here, we see that a 4 × 4 Hadamard matrix generates four orthogonal and four antipodal codes, for a total of eight bi-orthogonal codes as tabulated below:

| 4 Bit orthogonal code block | 4 Bit antipodal code block |
|---|---|
| 0 0 0 0 | 1 1 1 1 |
| 0 1 0 1 | 1 0 1 0 |
| 0 0 1 1 | 1 1 0 0 |
| 0 1 1 0 | 1 0 0 1 |

This principle can be extended to generate $n$ orthogonal codes and $n$-antipodal codes, for a total of 2n bi-orthogonal codes. Table below provides a few orthogonal and antipodal code sts for $n = 2, 4, 8, 16, 32, 64$.

| Code length ($n$) | Orthogonal codes ($n$) | Antipodal codes ($n$) | Bi-orthogonal codes ($2n$) |
|---|---|---|---|
| 2 | 2 | 2 | 4 |
| 4 | 4 | 4 | 8 |
| 8 | 8 | 8 | 16 |
| 16 | 16 | 16 | 32 |
| 32 | 32 | 32 | 64 |
| 64 | 64 | 64 | 128 |
| ⋮ | ⋮ | ⋮ | ⋮ |

## 5.4.2   Bi-orthogonal Codes

In the above, we have established that orthogonal codes are binary valued and have equal numbers of 1's and 0's. Antipodal codes, on the other hand, are just the inverse of orthogonal codes. Antipodal codes are also orthogonal among themselves. Therefore, an $n$-bit orthogonal code has $n$-orthogonal codes and $n$-antipodal codes, for a total of $2n$ bi-orthogonal codes. For example, an eight-bit orthogonal code has eight orthogonal codes and eight antipodal codes, for a total of 16 bi-orthogonal codes as shown in Fig. 5.8.

Similarly, a 16-bit orthogonal code has 16 orthogonal code and 16-antipodal code for a total of 32 bi-orthogonal codes, as shown in Fig. 5.9. We will take this bi-orthogonal codes block as an example and examine the error control properties.

**Fig. 5.8**  Bi-orthogonal code set for $n = 8$. An 8-bit orthogonal code has eight orthogonal code and eight antipodal code for a total of 16 bi-orthogonal codes

| Orthogonal Code | Antipodal Code |
|---|---|
| 0 0 0 0 0 0 0 0 | 1 1 1 1 1 1 1 1 |
| 0 1 0 1 0 1 0 1 | 1 0 1 0 1 0 1 0 |
| 0 0 1 1 0 0 1 1 | 1 1 0 0 1 1 0 0 |
| 0 1 1 0 0 1 1 0 | 1 0 0 1 1 0 0 1 |
| 0 0 0 0 1 1 1 1 | 1 1 1 1 0 0 0 0 |
| 0 1 0 1 1 0 1 0 | 1 0 1 0 0 1 0 1 |
| 0 0 1 1 1 1 0 0 | 1 1 0 0 0 0 1 1 |
| 0 1 1 0 1 0 0 1 | 1 0 0 1 0 1 1 0 |

| 16 Bit Orthogonal Code | 16 Bit Antipodal Code |
|---|---|
| 0 0 0 0 0 0 0 0 0 0 0 0 0 0 0 0 | 1 1 1 1 1 1 1 1 1 1 1 1 1 1 1 1 |
| 0 1 0 1 0 1 0 1 0 1 0 1 0 1 0 1 | 1 0 1 0 1 0 1 0 1 0 1 0 1 0 1 0 |
| 0 0 1 1 0 0 1 1 0 0 1 1 0 0 1 1 | 1 1 0 0 1 1 0 0 1 1 0 0 1 1 0 0 |
| 0 1 1 0 0 1 1 0 0 1 1 0 0 1 1 0 | 1 0 0 1 1 0 0 1 1 0 0 1 1 0 0 1 |
| 0 0 0 0 1 1 1 1 0 0 0 0 1 1 1 1 | 1 1 1 1 0 0 0 0 1 1 1 1 0 0 0 0 |
| 0 1 0 1 1 0 1 0 0 1 0 1 1 0 1 0 | 1 0 1 0 0 1 0 1 1 0 1 0 0 1 0 1 |
| 0 0 1 1 1 1 0 0 0 0 1 1 1 1 0 0 | 1 1 0 0 0 0 1 1 1 1 0 0 0 0 1 1 |
| 0 1 1 0 1 0 0 1 0 1 1 0 1 0 0 1 | 1 0 0 1 0 1 1 0 1 0 0 1 0 1 1 0 |
| 0 0 0 0 0 0 0 0 1 1 1 1 1 1 1 1 | 1 1 1 1 1 1 1 1 0 0 0 0 0 0 0 0 |
| 0 1 0 1 0 1 0 1 1 0 1 0 1 0 1 0 | 1 0 1 0 1 0 1 0 0 1 0 1 0 1 0 1 |
| 0 0 1 1 0 0 1 1 1 1 0 0 1 1 0 0 | 1 1 0 0 1 1 0 0 0 0 1 1 0 0 1 1 |
| 0 1 1 0 0 1 1 0 1 0 0 1 1 0 0 1 | 1 0 0 1 1 0 0 1 0 1 1 0 0 1 1 0 |
| 0 0 0 0 1 1 1 1 1 1 1 1 0 0 0 0 | 1 1 1 1 0 0 0 0 0 0 0 0 1 1 1 1 |
| 0 1 0 1 1 0 1 0 1 0 1 0 0 1 0 1 | 1 0 1 0 0 1 0 1 0 1 0 1 1 0 1 0 |
| 0 0 1 1 1 1 0 0 1 1 0 0 0 0 1 1 | 1 1 0 0 0 0 1 1 0 0 1 1 1 1 0 0 |
| 0 1 1 0 1 0 0 1 1 0 0 1 0 1 1 0 | 1 0 0 1 0 1 1 0 0 1 1 0 1 0 0 1 |

**Fig. 5.9**  Bi-orthogonal code set for $n = 16$. A 16-bit orthogonal code has 16 orthogonal code and 16-antipodal code for a total of 32 bi-orthogonal codes

Thus we conclude that an orthogonal code has the following properties:
Binary valued

- Has equal number of 1's and 0's
- Zero cross-correlation property

## 5.5 Orthogonal Codes for Data Spreading and De-spreading

In CDMA, each user is assigned a unique $n$-bit orthogonal code as a user ID, spectrum spreading at the transmit side, and dispreading at the receive side. Spectrum spreading is accomplished by multiplying each NRZ data bit by means of an $n$-bit orthogonal code. Multiplication in this process is referred to as exclusive OR (EXOR) operation. The output of the EXOR (Exclusive OR gate) is now a high-speed orthogonal or antipodal code. De-spreading is a similar process, where the receiver multiplies the incoming data by means of the same orthogonal code and recovers the data. Let's examine these operations using a 4-bit orthogonal code, where the code sequence is given by 0011 and the input NRZ data is 0 and 1.

### 5.5.1 Spreading Bit 0 and De-spreading to Recover Bit 0

This is shown in Fig. 5.10. When the binary bit 0 is multiplied by a 4-bit orthogonal code 0011, we write,

$$0 \text{ EXOR } (0011) = 0011 \tag{5.10}$$

This is the orthogonal code, reproduced due to exclusive OR operation. Moreover, the bit rate is also multiplied by a factor of four, thereby spreading the spectrum by a factor of four as well. This is the wide-band data which is transmitted to the receiver. Upon receiving 0011, the receiver performs the de-spreading function

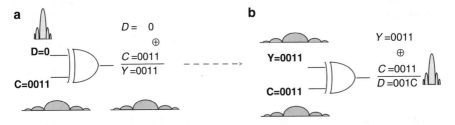

**Fig. 5.10** Spreading and de-spreading technique. (**a**) Spreading bit 0. (**b**) De-spreading and recovering bit 0

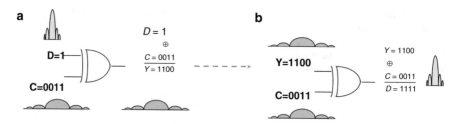

**Fig. 5.11** Spreading and de-spreading technique. (**a**) Spreading bit 1. (**b**) De-spreading and recovering bit 1

using the same orthogonal code 0011, which is also an EXOR function. Thus we write:

$$0011 \text{ EXOR } 0011 = 0000$$

This is the original data bit 0, which has been reproduced due to exclusive OR operation. Moreover, the bit rate is also divided by a factor of four.

### 5.5.2  Spreading Bit 1 and De-spreading to Recover Bit 1

This is shown in Fig. 5.11. When the binary bit 1 is multiplied by the same orthogonal code, we obtain,

$$1 \text{ EXOR } (0011) = 1100 \tag{5.11}$$

This is the inverse of the orthogonal code. This code is also known as antipodal code. The spectrum is also spread by a factor of four.

Upon receiving the antipodal code 1100, the receiver performs the de-spreading function using the same orthogonal code 0011, which is also an EXOR function. Thus we write:

11100 EXOR 0011 = 1111. This represents the original data Bit 1, which has been reproduced due to exclusive OR operation. Moreover, the bit rate is also divided by a factor of four.

### 5.5.3  Multiuser CDMA

In CDMA, each user is assigned a unique orthogonal code. Therefore, after de-spreading, all users, other than the desired user, will remain spread and will appear as noise. This is shown in Fig. 5.12 for two users.

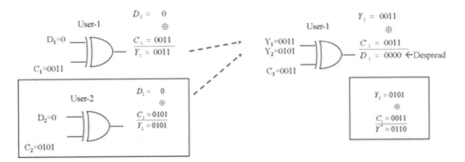

**Fig. 5.12** Multiuser CDMA. Each user is assigned a unique orthogonal code. Therefore, after de-spreading, all users will remain spread other than the desired recipient [5]

### 5.5.4  *Observations on Spreading and De-spreading*

- When an information bit is multiplied by an $n$-bit orthogonal code, the spectrum is spread by a factor of $n$.
- When the spread data is multiplied by the same orthogonal code, the spectrum is de-spread by the same factor, and the original data is recovered.
- When bit 0 is multiplied by an orthogonal code, the outcome is the same orthogonal code.
- When bit 1 is multiplied by an orthogonal code, the outcome is the antipodal code. The antipodal code is the inverse of the orthogonal code.
- When the spectrum is spread by a certain factor, the amplitude of the power spectral density reduces by the same factor. This is due to conservation of energy.
- This process is known as spread *spectrum*, resulting in a process gain, defined as: $G_s = 10\log(BW/R_b)$, where BW is the bandwidth and $R_b$ is the bit rate.
- Therefore, spread spectrum means spreading *the power spectrum over a predetermined band.*

## 5.6  Construction of CDMA Radio

### 5.6.1  *Background*

In CDMA (Code Division Multiple Access), multiple users have access to the same carrier frequency at the same time. A single-carrier frequency is used by several users, where each user is assigned a unique orthogonal code. The use of orthogonal codes enables each user to communicate without interference. CDMA is also known as spread spectrum technique. This is accomplished by multiplying each information bit by means of an $n$-bit orthogonal code. Multiplication in this process is referred to as exclusive OR (EXOR) operation.

For example, when the binary bit 0 is multiplied by a 4-bit orthogonal code 0101, we write,

$$0 \text{ EXOR } (0101) = 0101 \tag{5.12}$$

This is the orthogonal code that is reproduced due to exclusive OR operation. Moreover, the bit rate is also multiplied by a factor of four, thereby spreading the spectrum by a factor of four as well.

Similarly, when the binary bit 1 is multiplied by the same orthogonal code, we obtain,

$$1 \text{ EXOR } (0101) = 1010$$

This is the inverse of the orthogonal code. This code is also known as antipodal code. The spectrum is also spread by a factor of four. Therefore, when an information bit is multiplied by an $n$-bit orthogonal code, the spectrum is spread by a factor of $n$, enabling multiple users to share the same spectrum at the same time. For this reason a large bandwidth is assigned for CDMA radio.

### 5.6.2  Binary Bit 0 Transmit/Receive Mechanism

Let's examine a CDMA radio based on 4-bit orthogonal code, where the code sequence is given by 0101. This is shown in Figs. 5.13 and 5.14.

In Fig. 5.13, the binary bit 0 is multiplied by the 4-bit orthogonal code to reproduce the orthogonal code as follows:

$$0 \text{ EXOR } (0101) = 0101$$

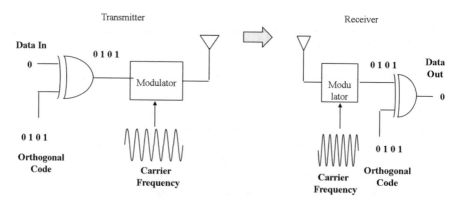

**Fig. 5.13**  CDMA radio. Binary bit 0 transmit/receive mechanism

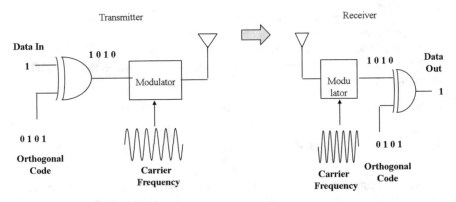

**Fig. 5.14** CDMA radio. Binary bit 1 transmit/receive mechanism

This represents the information bit 0, which is modulated and transmitted to the receiver.

The receiver intercepts the modulated carrier frequency, demodulates and recovers the orthogonal code 0101. Since the exclusive OR gate also uses the same orthogonal code, we obtain,

$$0101 \text{ EXOR } (0101) = 0000 \tag{5.13}$$

This represents the original binary value 0.

### 5.6.3    Binary Bit 1 Transmit/Receive Mechanism

Similarly, in Fig. 5.14, the binary bit 1 is multiplied by the same 4-bit orthogonal code to produce the antipodal code as follows:

$$1 \text{ EXOR } (0101) = 1010$$

This represents the information bit 1, which is modulated and transmitted to the receiver.

The receiver intercepts the modulated carrier frequency, demodulates and recovers the antipodal code 1010. Since the exclusive OR gate uses the same orthogonal code, we obtain,

$$1010 \text{ EXOR } (0101) = 1111 \tag{5.14}$$

This represents the original binary value 1.

In summary, CDMA is a branch of multiple access technique, where multiple users have access to the same spectrum through orthogonal codes. Orthogonal codes are binary valued and have equal number of 1's and 0's. Therefore, for an *n*-bit orthogonal code, there are *n*-orthogonal codes. In CDMA, each user is assigned a unique orthogonal code. As a result, each user remains in orthogonal space after modulation and offers maximum isolation. Yet, there is a limit to the use of all the codes, which is related to channel capacity. We will address this topic next.

## 5.7  CDMA Capacity

Figure 5.15 illustrates the CDMA capacity. Here, each user appears as noise due to spreading. Therefore, the interference is due to all users except one who is dispreading. It follows that the total interference will be given by:

$$I = C(N - 1)$$

or

$$\frac{C}{I} = \frac{1}{N - 1} \tag{5.15}$$

We also note that the carrier-to-interference ratio is generally described by:

$$\frac{C}{I} = \frac{R_b \times E_b}{N_o \times W} \tag{5.16}$$

**Fig. 5.15** Illustration of CDMA capacity. Each user occupies a bandwidth W after spreading. One who is de-spreading has a process gain to recover the data [5]

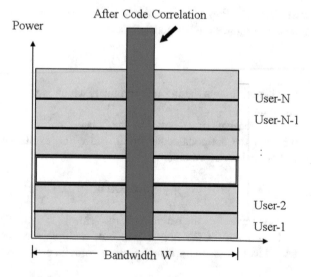

where:

- $E_b$ = energy per bit
- $R_b$ = bit rate
- $N_o$ = thermal noise
- $W$ = bandwidth after spreading

Equating (5.15) and (5.16) we get

$$\frac{1}{N-1} = \frac{R_b \times E_b}{N_o \times W} \tag{5.17}$$

Solving for the CDMA capacity $N$, we obtain,

$$N = 1 + \frac{W/R_b}{E_b/N_o} \tag{5.18}$$

In the above capacity equation, the bandwidth after spreading relates to orthogonal code. Therefore, we can write,

- $W = 2R_c$
- $R_c$ = code rate  $\qquad$ (5.19)

The CDMA capacity then becomes,

$$N = 1 + \frac{2R_c + R_b}{E_b/N_o}$$

In terms of code length $n$, the capacity equation can be further modifies as,

$$N = 1 + \frac{2n}{E_b/N_o} \tag{5.20}$$

In the above equation, $n$ is the code length, and $N$ is the total number of users that can share the same bandwidth, using an $n$-bit orthogonal code. Table 5.2 shows the estimates of CDMA capacity for various code lengths.

It may be noted that the process gain alone is not responsible for the overall system performance and capacity. Other factors such as error control coding, cell site

**Table 5.2** CDMA capacity for various code lengths

| Code length ($n$) | $E_b/N_o$ (dB) | $E_b/N$ (linear) | CDMA capacity ($N$) |
|---|---|---|---|
| 4 | 6 | 3.98107171 | 3.009509 |
| 8 | 6 | 3.98107171 | 5.019018 |
| 16 | 6 | 3.98107171 | 9.038037 |
| 32 | 6 | 3.98107171 | 17.07607 |
| 64 | 6 | 3.98107171 | 33.15215 |

deployment, antenna directivity and down tilt, sectorization, frequency reuse, etc. play important roles in enhancing the overall performance and capacity of the CDMA system.

## 5.8   Conclusions

This chapter presents a brief overview of spread spectrum CDMA radio. Spreading and dispreading techniques are described with illustrations to show how it relates to spread spectrum CDMA radio. Orthogonal codes and their properties are presented and show how Orthogonal Codes are generated and used to design CDMA radio. CDMA channel capacity is derived as a function of code length. Since CDMA is a noise-limited system, the need for power control is described to resolve the near-far problems.

## References

1. IS-95. (1993, March 15). *Mobile station – Base station compatibility standard for dual mode wide band spread spectrum cellular systems* (TR 55, PN-3115).
2. Qualcomm. (1993). *The CDMA network engineering handbook: Concepts in CDMA* (Vol. 1). Qualcom, California, U.S.A.
3. Yang, S. C. (1998). *CDMA RF system engineering.* Norwood, MA: Artech House.
4. Garg, V. K. (1999). *IS-95 CDMA and cdma 2000.* Upper Saddle River, NJ: Prentice Hall.
5. Faruque, S. (1996). *Cellular mobile systems engineering.* Boston: Artech House.

# Chapter 6
# Orthogonal Frequency Division Multiple Access (OFDMA)

**Abstract** This chapter presents a brief overview of the OFDMA technique used in 4G WiMAX (Worldwide Interoperability for Microwave Access) and 4G LTE (Long-Term Evolution) cellular system. It is shown that OFDMA is an extension of FDMA, where each frequency band is placed at the null of the adjacent frequency band. This is governed by the well-known "Fourier transform," so that adjacent frequency bands are orthogonal to each other. OFDMA is a full-duplex communication system. The communication link is maintained in both directions in the time domain known as time division duplex (TDD). Numerous illustrations are used to bring students up to date in key concepts, underlying principles, and practical applications of "Fourier transform," spectrum, and orthogonal properties of spectrum, leading to OFDMA. Construction of OFDMA channels from the FCC-allocated band is presented to illustrate the concept.

**Objectives**

- Introduce OFDMA and show how it relates to FDMA.
- Review Fourier transform and derive spectral response of data.
- Recognize orthogonal signals and derive OFDMA channels from FCC-allocated frequency band.
- Design OFDMA radio and show spectrum allocation scheme.
- Review TDD and show how TDD is used in OFDMA.

## 6.1 Introduction to OFDMA

Multiple access technique is well known in cellular communications [1–6]. It enables many users to share the same spectrum in the frequency domain, time domain, code domain, or phase domain. It begins with a frequency band, allocated by FCC (Federal Communications Commission) [8]. FCC provides licenses to operate wireless communication systems over given bands of frequencies. These

© The Author(s), under exclusive license to Springer Nature Switzerland AG. 2019     63
S. Faruque, *Radio Frequency Multiple Access Techniques Made Easy*,
SpringerBriefs in Electrical and Computer Engineering,
https://doi.org/10.1007/978-3-319-91651-4_6

bands of frequencies are finite and have to be further divided into smaller bands (channels) and reused to provide services to other users. This is governed by the International Telecommunication Union (ITU) [9]. ITU generates standards such as FDMA, TDMA, CDMA, OFDMA, etc. for wireless communications.

Figure 6.1 illustrates the role of FCC-allocated spectrum and shows how OFDMA is derived from FDMA.

*In FDMA* (frequency division multiple access) [1–3], a carrier frequency is assigned to a single user. In this scheme, the channel is occupied by a single user for the entire duration of the call. FDMA is the multiple access technique used in AMPS (Advanced Mobile Phone System), also known as first-generation (1G) cellular communication system.

*OFDMA* (orthogonal frequency division multiple access) [6, 7] is an extension of FDMA, where each frequency is placed at the null of the adjacent frequency (see Fig. 6.1). This is governed by the well-known "Fourier transform," so that adjacent frequencies are orthogonal to each other. It begins with a band of frequencies. This band of frequencies is allocated by the FCC (Federal Communications Commission). This band of frequency is further divided into several narrow bands of frequencies, where each frequency is orthogonal to each other. OFDMA is a full-duplex communication system. The communication link is maintained in both directions in the time domain known as time division duplex (TDD).

OFDMA is a relatively new wireless communication standard used in 4G WiMAX (Worldwide Interoperability for Microwave Access) and 4G LTE (Long-Term Evolution) protocol. It may be noted that WiMAX is an IEEE 802.16 standard, while LTE is a standard developed by the 3GPP group. Both standards are surprisingly similar and bandwidth efficient. OFDMA is used in 4G cellular standard.

In any multiple access technique, multiple users have access to the same spectrum, so that the occupied bandwidth does not exceed the FCC-allocated channel. Furthermore, as the size and speed of digital data networks continue to expand, bandwidth efficiency becomes increasingly important. This is especially true for broadband communication, where the choice of modulation schemes is important keeping in mind the available bandwidth resources, allocated by the FCC. With these constraints in mind, this chapter will present a brief overview of the OFDMA technique used in 4G cellular communication systems. Numerous illustrations will

**Fig. 6.1** Illustration of FCC-allocated spectrum showing how OFDMA is derived from FDMA

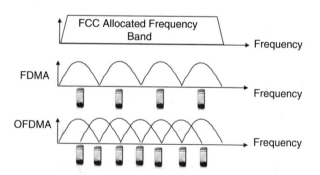

be used to bring students up to date in key concepts, underlying principles, and practical applications of "Fourier transform," spectrum, and orthogonal properties of spectrum, leading to OFDMA. Construction of OFDMA channels from FCC-allocated band is presented to illustrate the concept.

## 6.2 Spectral Response of Data Before Modulation

In digital communications, data is generally referred to as a nonperiodic digital signal. It has two values:

- Binary-1 = High, Period = $T$
- Binary-0 = Low, Period = $T$

Also, data can be represented in two ways:

- Time domain representation
- Frequency domain representation

The time domain representation of data (Fig. 6.2a), known as non-return-to-zero (NRZ) data, is given by:

$$V(t) = V \quad < 0 < t < T$$
$$= 0 \quad \text{elsewhere} \tag{6.1}$$

The frequency domain representation is given by "Fourier transform [xx]":

$$V(\omega) \int_0^T V.e^{-j\omega t} dt \tag{6.2}$$

$$| V(\omega) | = VT \left[ \frac{\sin(\omega T/2)}{\omega T/2} \right] \tag{6.3}$$

$$P(\omega) = \left( \frac{1}{T} \right) |V(\omega)|^2 = V^2 T \left[ \frac{\sin(\omega T/2)}{\omega T/2} \right]^2 \tag{6.4}$$

Here, $P(\omega)$ is the power spectral density. This is plotted in Fig. 6.2b. The main lobe corresponds to the fundamental frequency, and side lobes correspond to harmonic components. The bandwidth of the power spectrum is proportional to the frequency. In practice, the side lobes are filtered out since they are relatively insignificant with respect to the main lobe. Therefore, the one-sided bandwidth is

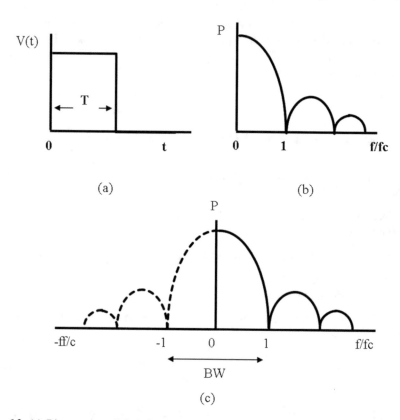

**Fig. 6.2** (**a**) Discrete time digital signal and (**b**) its one-sided power spectral density and (**c**) two-sided power spectral density. The bandwidth associated with the non-return-to-zero (NRz) data is $2R_b$, where $R_b$ is the bit rate

given by the ratio $f/f_b = 1$. In other words, the one-sided bandwidth $= f = f_b$, where $f_b = R_b = 1/T$, $T$ being the bit duration.

The general equation for two-sided response is given by:

$$V(\omega) = \int_{-\infty}^{\infty} V(t).e^{-j\omega t} dt \qquad (6.5)$$

In this case, $V(\omega)$ is called two-sided spectrum of $V(t)$. This is due to both positive and negative frequencies used in the integral. The function can be a voltage or a current. Figure 6.2c shows the two-sided response, where the bandwidth is determined by the main lobe.

Notice the two nulls before and after the main lobe. These nulls are the location of adjacent channels. These nulls are also recognized as "orthogonal" frequencies as we shall see it later. We will now take ASK and BPSK modulation schemes to examine the shifting properties of spectral response of data after modulation.

## 6.3   Spectral Response of Data After Modulation

Spectral response of data after modulation depends on the bit rate and the modulation type [10–12]. We will examine this by means of NRZ data and MPSK modulation with $M = 2, 4, 8$, etc. and evaluate the corresponding transmission bandwidth and OFDMA spectrum.

### 6.3.1   Spectral Response After BPSK Modulation

In BPSK, the phase of the carrier changes in two discrete levels, in accordance to the input signal. We can represent this by means of a signal constellation diagram with $M = 2$, 1 bit per phase. This is shown in Fig. 6.3, where the input raw data, having a bit rate Rb1, is encoded by means of a rate r encoder. The encoded data, having a bit rate $R_{b2} = R_{b1}/r\ (r < 1)$, is modulated by the BPSK modulator as shown in the figure.

The BPSK modulator takes one bit at a time to construct the phase constellation having two phases, also known as "symbols," where each symbol represents 1 bit. The symbol rate is therefore the same as the encoded bit rate $R_{b2}$.

Therefore, the BPSK modulator has the following specifications:

- Two phases or two symbols
- 1 bit/symbol

Figure 6.4 shows the spectral response of the BPSK modulator. Since there are two phases, the carrier frequency changes in two discrete levels, 1 bit per phase, as

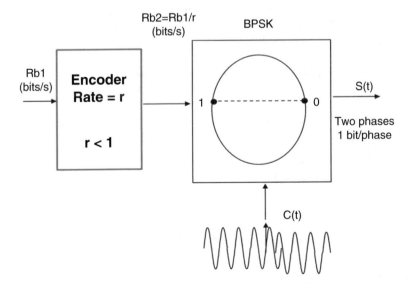

**Fig. 6.3**  BPSK signal constellation having two symbols, 1 bit per symbol

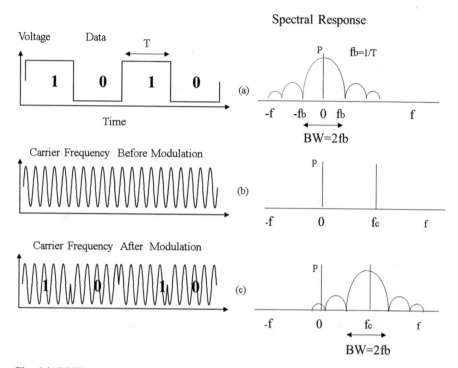

**Fig. 6.4** BPSK spectrum at a glance. (**a**) Spectral response of NRZ data before modulation. (**b**) Spectral response of the carrier before modulation. (**c**) Spectral response of the carrier after modulation

shown in the figure. Notice that the spectral response after BPSK modulation is the shifted version of the NRZ data, centered on the carrier frequency $f_c$.

The transmission bandwidth is given by

$$\begin{aligned}
\mathrm{BW(BPSK)} &= 2R_2/\text{bits per symbol} \\
&= 2R_{b2}/1 = 2R_{b2} = 2R_{b1}/r \ \ \mathrm{Hz}
\end{aligned} \tag{6.6}$$

This is the OFDMA channel bandwidth for BPSK modulation, where:

- $R_{b1}$ is the bit rate before coding.
- $R_{b2}$ = bit rate after coding = $R_{b1}/r$.
- $r$ = code rate ($r = 1/2$ typical).
- Bits per symbol =1 (BPSK: $\varphi = 2$).

### 6.3.2   Spectral Response After QPSK Modulation

In QPSK, the input raw data, having a bit rate $R_{b1}$, is encoded by a rate $r$ ($r < 1$) encoder. The encoded data, having a bit rate $R_{b2} = R_{b1}/r$, is serial to parallel

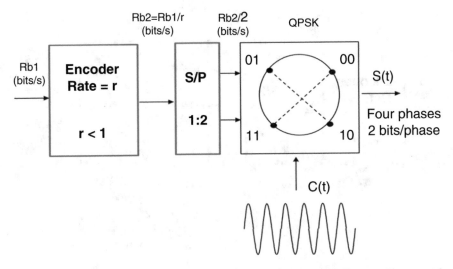

**Fig. 6.5** QPSK signal constellation having four symbols, 2 bits per symbol

converted into two parallel bit streams. The encoded bit rate, now reduced in speed by a factor of two, is modulated by the QPSK modulator as shown in Fig. 6.5.

The QPSK modulator takes one bit from each stream to construct the phase constellation having four phases, also known as "symbols," where each symbol represents 2 bits. The symbol rate is therefore reduced by a factor of two. The QPSK modulator has four phases or four symbols, 2 bits/symbol, as shown in the figure.

Therefore, the QPSK modulator has the following specifications:

- Four phases or four symbols
- 2 bits/symbol

The transmission bandwidth is given by

$$\begin{aligned}
\mathrm{BW(QPSK)} &= 2R_2/\text{bits per symbol} \\
&= 2R_{b2}/2 = R_{b2} = R_{b1}/r \ \text{Hz}
\end{aligned} \tag{6.7}$$

This is the OFDMA channel bandwidth for QPSK modulation, where:

- $R_{b1}$ is the bit rate before coding.
- $R_{b2} = $ bit rate after coding $= R_{b1}/r$.
- $r = $ code rate ($r = 1/2$ typical).
- Bits per symbol $= 2$ (QPSK: $\varphi = 4$).

### 6.3.3   Spectral Response After 8PSK Modulation

In 8PSK, the input raw data, having a bit rate $R_{b1}$, is encoded by a rate $r$ ($r < 1$) encoder. The encoded data, having a bit rate $R_{b2} = R_{b1}/r$, is serial to parallel

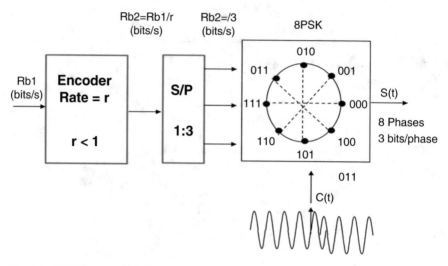

**Fig. 6.6** 8PSK signal constellation having eight symbols, 3 bits per symbol

converted into three parallel streams. The encoded bit rate, now reduced in speed by
a factor of three, is modulated by the 8PSK modulator as shown in Fig. 6.6.

The 8PSK modulator takes one bit from each stream to construct the phase
constellation having 8 phases, also known as "Symbols", where each symbol
represents 3 bits. The symbol rate is therefore reduced by a factor of 3. The 8PSK
modulator has 8 phases or 8 symbols, 3 bits/symbol, as shown in the figure.

Therefore, the 8PSK modulator has the following specifications:

- Eight phases or eight symbols
- 3 bits/symbol

The transmission bandwidth is given by

$$\begin{aligned} BW(8PSK) &= 2R_2/\text{bits per symbol} \\ &= 2R_{b2}/3 = (2/3)R_{b2} = (2/3)R_{b1}/r \ \ \text{Hz} \end{aligned} \tag{6.8}$$

This is the OFDMA channel bandwidth for 8PSK modulation, where:

- $R_{b1}$ is the bit rate before coding.
- $R_{b2} =$ bit rate after coding $= R_{b1}/r$.
- $r =$ code rate ($r = 1/2$ typical).
- Bits per symbol $= 3$ (8PSK: $\varphi = 8$).

### 6.3.4 Spectral Response After 16PSK Modulation

In 16PSK, the input raw data, having a bit rate $R_{b1}$, is encoded by means of a rate
$r$ ($r < 1$) encoder. The encoded data, having a bit rate $R_{b2} = R_{b1}/r$, is serial to parallel

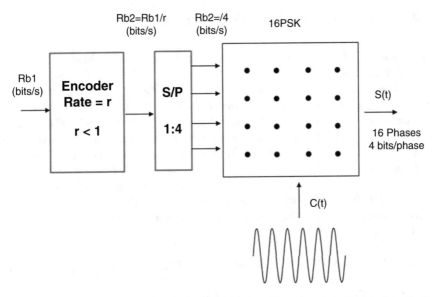

**Fig. 6.7** 16 PSK signal constellation having 16 symbols, 4 bits per symbol. Here, each symbol is represented by a dot, where each dot represents 4 bits

converted into four parallel streams. The encoded bit rate, now reduced in speed by a factor of four, is modulated by the 16PSK modulator as shown in Fig. 6.7.

The 16PSK modulator takes 1 bit from each stream to construct the phase constellation having 16 phases, also known as "symbols," where each symbol represents four bits. The symbol rate is therefore reduced by a factor of four. Therefore, the 16PSK modulator has 16 phases or 16 symbols, 4 bits/symbol, as shown in the figure.

Therefore, the 16PSK modulator has the following specifications:

- 16 phases or 16 symbols
- 4 bits/symbol

The transmission bandwidth is given by

$$\begin{aligned} BW(16PSK) &= 2R_2/\text{bits per symbol} \\ &= 2R_{b2}/4 = (1/2)R_{b2} = (1/2)R_{b1}/r \ \text{Hz} \end{aligned} \tag{6.9}$$

This is the OFDMA channel bandwidth for 16PSK modulation, where:

- $R_{b1}$ is the bit rate before coding.
- $R_{b2} =$ bit rate after coding $= R_{b1}/r$.
- $r =$ code rate ($r = 1/2$ typical).
- Bits per symbol $= 4$ (8PSK: $\varphi = 16$).

## 6.4  Construction of OFDMA Channels

Stated earlier, OFDMA channels are constructed from FDMA channels. It begins with a wide frequency band, allocated by the FCC (Federal Communications Commission). In this scheme, shown in Fig. 6.8, the FCC-allocated frequency band is divided into several smaller frequency bands known as FDMA channels. The bandwidth of each FDMA channel is determined by the main lobe of the power spectral density, which is governed by the input coded bit rate and the modulation type. The bandwidth is defined by the first null to first null of the two-sided response of the main lobe.

In OFDMA, each frequency band is placed at the null of the adjacent band as shown in Fig. 6.8. Notice that there are $n - 1$ nulls in the OFDMA spectrum, where $n$ is the number of FDMA channels. Therefore, the number of OFDMA channels will be given by

$$\text{Number of OFDMA channels} = 2 \times \text{FDMA channels} - 1 \qquad (6.10)$$

Let's examine this by means of the following problems:

**Problem 1**
In this problem we will construct OFDMA channels from a given FCC- allocated frequency band with the following design parameters:

Given:
- FCC-allocated band = 20 MHz
- Input bit rate before coding $R_{b1}$ = 1 MHz
- Rate ½ convolutional Coding ($r = 1/2$)
- QPSK modulationFind:
- Bit rate after coding $R_{b2}$

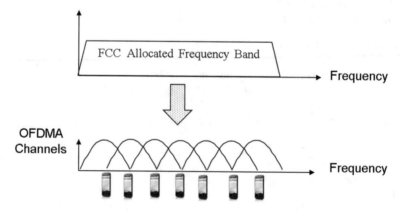

**Fig. 6.8** Construction of OFDMA channels from FCC-allocated frequency band. There are $n - 1$ nulls in the FCC-allocated band, where $n$ is the number of FDMA channels

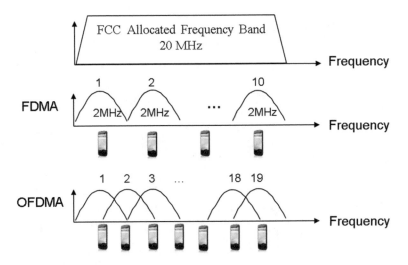

**Fig. 6.9**  Construction of FDMA and OFDMA channels

- The required channel bandwidth BW
- Number of FDMA channels available
- Number of OFDMA channels available

Solution:
- Bit rate after coding $R_{b2} = R_{b1}/r = 2R_{b1} = 2 \times 1$ Mbps = 2 Mbps.
- BW(QPSK) $= 2R_{b2}/$bits per symbol = 4 Mbps/2 = 2 MHz.
- Number of FDMA channels = 20 MHz/BW = 20 MHz/2 MHz = 10 FDMA channels (see figure below).
- Number of OFDMA channels = 2FDMA channels $- 1 = 20 - 1 = 19$ OFDMA channels (see figure below).
- Figure 6.9 illustrates the channel sets.

**Problem 2**
Repeat the previous problem with 16 PSK modulation.

Given:
- FCC-allocated band = 20 MHz
- Input bit rate before coding $R_{b1} = 1$ MHz
- Rate ½ convolutional coding ($r = 1/2$)
- 16PSK modulationFind:
- Bit rate after coding $R_{b2}$
- The required channel bandwidth BW
- Number of FDMA channels available
- Number of OFDMA channels available

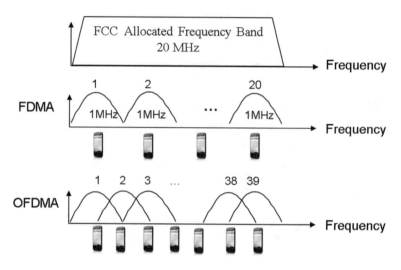

**Fig. 6.10**  Construction of FDMA and OFDMA channels with 16PSK modulation

Solution:
- Bit rate after coding $R_{b2} = R_{b1}/r = 2R_{b1} = 2 \times 1$ Mbps $= 2$ Mbps.
- BW(16PSK) $= 2R_{b2}$/bits per symbol $= 4$ Mbps/4 $= 1$ MHz.
- Number of FDMA channels $=20$ MHz/BW $= 20$ MHz/1 MHz $= 20$ FDMA channels (see figure below).
- Number of OFDMA channels $= 2$FDMA channels $- 1 = 40-1 = 39$ OFDMA channels (see figure below).
- Figure 6.10 illustrates the channel sets.

## 6.5  OFDMA-TDD Mode of Operation

OFDMA is a full-duplex communication system. The communication link is maintained in both directions in the time domain known as time division duplex (TDD). In OFDMA-TDD, a single frequency is time shared between the uplink and the downlink. The duration of transmission in each direction is generally short, in the order of ms. In this scheme, when the mobile transmits, base station listens, and when the base station transmits, mobile listens. This is accomplished by formatting the data into a "frame," where the frame is a collection of several time slots. Each time slot is a package of data, representing digitized voice, digitized text, digitized video, and synchronization bits (sync. bits). The sync bits are unique, which are used for frame synchronization.

Figure 6.11 illustrates a typical frame and the TDD transmission scheme.

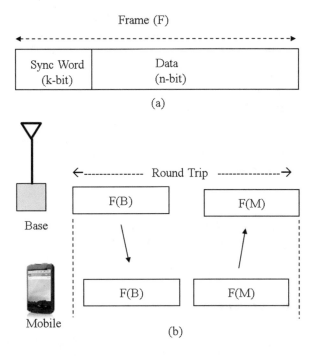

**Fig. 6.11** OFDMA-TDD frame structure. (**a**) Frame. (**b**) Frame transmission scheme

According to TDD transmission, both the base station and the mobile use the same carrier frequency. The steps involved in transmit/receive mechanism between the base station and mobile are as follows:

- The base station modulates the carrier frequency by means of the digital information bits in frame $F(B)$ and transmits to the mobile.
- Since the mobile is tuned to the same carrier frequency, it receives the frame $F(B)$ after a propagation delay $t_\mathrm{P}$.
- Mobile synchronizes the frame using the sync bits and downloads the data.
- After a guard time $t_\mathrm{g}$, mobile transmits its own frame $F(M)$ to the base using the same carrier frequency.
- Base receives the frame from the mobile after a propagation delay $t_\mathrm{p}$, maintains sync using the sync bits, and downloads the data.
- A round-trip communication is now complete.
- The communication continues until one terminates the call.

As can be seen in the figure, the TDD schemes require a propagation delay and a guard time between transmission and reception. The complete round-trip delay $T_\mathrm{d}$ must be sufficient to accommodate the frame, the propagation delay, and the guard time. Therefore, the round-trip delay can be written as

$$T_\mathrm{d} = 2\big(F + t_\mathrm{p} + t_\mathrm{g}\big) \tag{6.11}$$

The round-trip delay $T_d$ depends on the frame length $F$, which is generally in milliseconds (ms). The propagation delay $t_p$ depends on the propagation distance, and the guard time $t_g$ depends on the technology

**Problem 3**

Given:

- Frame length $= 2$ ms
- Guard time $t_g = 0.01$ ms
- Distance between the base station and mobile $= 1$ km
- Velocity of light $c = 3 \times 10^8$ m/sFind:
- The round-trip delay $t_p$.

Solution:

- Propagation delay for a distance of 1 km $= 1$ km/$(3 \times 108$ m$) = 3.3$ ms
- Round-trip delay td $= 2$ (tp + tf + tg) $= 2(3.3$ ms $+ 2$ ms $+ 0.01$ ms$) = 5.31$ ms

## 6.6   Conclusions

This chapter presents a brief overview of the OFDMA technique used in 4G WiMAX (Worldwide Interoperability for Microwave Access) and 4G LTE (Long-Term Evolution) cellular system. It is shown that OFDMA is an extension of FDMA, where each frequency band is placed at the null of the adjacent band. This is governed by the well-known "Fourier transform," so that adjacent bands are orthogonal to each other. OFDMA is a full-duplex communication system. The communication link is maintained in both directions in the time domain known as time division duplex (TDD). Numerous illustrations are used to bring students up to date in key concepts, underlying principles, and practical applications of "Fourier transform," spectrum, and orthogonal properties of spectrum, leading to OFDMA. Construction of OFDMA channels from FCC-allocated band is presented to illustrate the concept.

## References

1. IS-54. (1989, December). *"Dual-mode mobile station-base station compatibility standard* (EIA/TIA Project Number 2215).
2. IS-95. (1993, March 15). "Mobile station – Base station compatibility standard for dual mode wide band spread spectrum cellular systems (TR 45, PN-3115).
3. MacDonald, V. H. (1969, January). The cellular concept. *The Bell System Technical Journal, 58* (1).
4. Lee, W. C. Y. (1989). *Mobile cellular telecommunications systems.* New York: McGraw-Hill.
5. Faruque, S. (1996). *Cellular mobile systems engineering.* Norwood, MA: Artech House. ISBN: 0-89006-518-6.

6. ITU-R. *Requirements related to technical performance for IMT-advanced radio interface(s)* (Report M.2134), Approved in November 2008.
7. Parkvall, S., & Astely, D. (2009, April). The evolution of LTE toward LTE Advanced. *Journal of Communications, 4*(3), 146–154. https://doi.org/10.4304/jcm.4.3.146-154.
8. FCC, Federal Communications Commission, Washington, DC.
9. ITU, International Telecommunications Union, Paris, France.
10. Faruque, S. (2015). *Radio frequency source coding made easy, Springer Briefs in Electrical and Computer Engineering* (1st ed.). ISBN-13: 968-3319156088, ISBN-10: 331915608.
11. Faruque, S. (2016). *Radio frequency channel coding made easy, Springer Briefs in Electrical and Computer Engineering* (1st ed.). ISBN-13: 968-3319211695, ISBN-10: 3319211692.
12. Faruque, S. (2016). *Radio frequency modulation made easy, Springer Briefs in Electrical and Computer Engineering* (1st ed.). ISBN-13: 968-3319412009, ISBN-10: 3319412000.

Printed in the United States
By Bookmasters